CATACLYSMS AND EARTH HISTORY

For Shelley

CATACLYSMS AND EARTH HISTORY

The Development of Diluvialism

RICHARD HUGGETT

CLARENDON PRESS · OXFORD
1989

Oxford University Press, Walton Street, Oxford OX2 6DP
Oxford New York Toronto
Delhi Bombay Calcutta Madras Karachi
Petaling Jaya Singapore Hong Kong Tokyo
Nairobi Dar es Salaam Cape Town
Melbourne Auckland
and associated companies in
Berlin Ibadan

Oxford is a trade mark of Oxford University Press

Published in the United States
by Oxford University Press, New York

British Library Cataloguing in Publication Data
Huggett, Richard 1929–
Cataclysms and earth history
1. Floods. Theories, history
I. title
551.48'9
ISBN 0–19–828713–5

Library of Congress Cataloging-in-Publication Data
Huggett, Richard J.
Cataclysms and earth history: the development of diluvialism /
Richard Huggett.
Bibliography: p.
Includes index.
1. Deluge. 2. Bible and geology. 3. Floods. 4. Geomorphology.
I. Title.
QE507.H84   1989   904'.5—dc20   89–8834
ISBN 0–19–828713–5

Set by Hope Services, Abingdon
Printed in Great Britain by
Bookcraft Ltd,
Midsomer Norton, Avon

Preface

I am a geographer with a special interest in Earth surface processes. I have for a long time felt uneasy about the methodology of physical geography. I believe that it is too constrained by a desire to explain everything in terms of processes observed at present acting a little more intensely in the past. A few years ago my attention was alerted to the possibility of global cataclysms when I stumbled across two books: Hugh Auchincloss Brown's *Cataclysms of the Earth* (1967), and Victor Clube and Bill Napier's *The cosmic serpent* (1982). In these two books, I discovered that there are two ways of creating cataclysms which, although recognized by early geologists, had generally been deemed improbable—pole shift and meteorite bombardment. From there I scoured the shelves of the library seeking out further information on these topics. It soon became clear to me that the enormous flood waves which either fast pole shift or meteorite bombardment might produce would, in spilling over continents, or continental margins at least, create some of the deposits and topographical features attributed by the old school of diluvialists either to the Noachian Flood or to a series of grand floods. I became hooked on the subject and decided to explore it as fully as possible. This book is the result of that exploration. I felt it worth committing my findings to print because, as far as I know, no existing book draws together the varied aspects of diluvialism. There are, of course, many books on the history of geology and geomorphology, but none of them deals exclusively with the development of diluvialism, and none of them traces the varied threads of diluvialist thought through to the present.

Some Earth scientists may be aware of the historical development of diluvialism, but I suspect that many, like me a few years ago, think of diluvialism as an historical curiosity, as a school of thought which perished some time early in the last century, just after the advent of uniformitarianism, when 'proper' Earth science commenced. The aim of the present book is to bring to the notice of Earth scientists at large, rather than to historians of geology and geomorphology, the rich and colourful history of diluvialist thought, from the green but scintillating ideas of classical poets and philosophers, through the dazzling debates of the Renaissance scholars, the glittering and lurid theories of the Restoration cosmogonists, the brilliant philosophies expounded during the halcyon years of the Enlightenment, and the shimmering rainbow of views which arose during the early nineteenth century, to the present redevelopment of diluvialism.

I make no claim to comprehensiveness. I have by no means exhausted the references to, or even the aspects of, the topics I discuss. My aim is simply to paint a broad canvas of diluvialist thought at various places and at different times, and to show, in a general way, how these ideas have evolved, one from another. I do, however, provide a good number of references from which a more prolonged attack on individual topics could be mounted by an interested party. Nor do I claim to have read all the pre-1800 books and papers referred to. Many of these early works are difficult to obtain and, in any case, would require translation. To have followed up these more obscure tracts would, I felt, have been unproductive given the nature of the book. Instead, I have relied on the excellent summaries of the early works given in the many histories of geology and geomorphology which are readily available. Full credit to these sources is given in the text. Unless otherwise noted, all italicized words and phrases in quotations are found in the original works. All quotations are reprinted in their original form, though generally the capitalization of letters has been dropped.

I should like to thank a number of people without whose help this book would have never seen the light of day. Firstly, I would thank Professor Ian Douglas for his continued faith and support in my rather off-beat research (unlike most research today, it involved little money and just a pen—well, word processor—and paper). Secondly, I would thank Andrew Schuller and Oxford University Press for getting the book into print. Thirdly, I would thank Graham Bowden and Nick Scarle, cartographers in the School of Geography at Manchester University, for drawing the diagrams. And, lastly, but mostly, I would thank my wife for endless cups of tea and endless patience.

R.J.H.

Manchester
8 August 1988

Contents

List of Figures

List of Tables

EDWARD. SEA-SHELLS, did you say, mother, in the heart of solid rocks, and far inland? There must surely be some mistake in this; at least, it appears to me incredible.

. . . . .

MRS. R. The history of the shells, my dear, and many other things no less wonderful, is contained in the science called GEOLOGY, which treats of the first appearance of rocks, mountains, valleys, lakes, and rivers; and the changes they have undergone, from the Creation and the Deluge, till the present time.

CHRISTINA. I always thought that the lakes, mountains, and valleys, had been created from the first by God, and that no further history could be given of them.

MRS. R. True, my dear; but yet we may without presumption, inquire into what actually took place at the creation; and, by examining stones and rocks as we now find them, endeavour to trace what changes they have undergone in the course of ages.

. . . . .

EDWARD. This will, indeed be romantic and interesting.

(Penn 1828: 1–8)

1

Introducing Diluvialism
Cataclysms, debacles, and deluges

On definitions

Diluvial, diluvium, diluvialism, diluvialist—these words are used
repeatedly throughout the following pages. They are, if not exactly
open to gross misinterpretation, then certainly a little ambiguous,
and it is wise to say something about the senses in which they will
be employed. A few other words—cataclysm, debacles, deluge,
flood, and inundation—are more or less equivocal, but it may pay
to spell out their meanings.

The word diluvial, or diluvian, is an adaptation of the Latin
diluvialis meaning of, or brought about by, a flood. It may refer to
any flood, but is commonly used in connection with the Noachian
Flood recorded in Genesis. In geology, the diluvial theory explains
certain of the Earth's surface features by reference to a general
flood, or to bouts of catastrophic action of waters. Diluvialists
explain such superficial geological features as boulder clay,
abraded and polished rock surfaces, and ossiferous gravels by the
hypothesis of a universal flood, or an extraordinary movement of
the waters of the globe. In other words, they 'ascribe to a universal
deluge such superficial deposits as they cannot readily reconcile
with the ordinary operations of water now going on' (Page 1865:
176). Diluvialists, in the general sense of the word, are students of
global or regional floods; in the more restricted sense of the word,
they are students of the Noachian Flood. Diluvialism, or diluvianism,
is the system of Earth history which attributes certain superficial
features to one or more universal floods. From the same root as
diluvial comes the word diluvium and its very rare equivalent,
diluvion. Both words are applied to those superficial deposits
which appear not to have been formed by the ordinary action of
water, but to have resulted from some extraordinary action of
water on a huge scale. Originally, the term diluvium was applied to
superficial deposits thought to be due to the Noachian Flood, but it

came generally to be 'applied to all masses apparently the result of powerful aqueous agency' (Page 1865: 176).

The word cataclysm is charged with emotion. It conjures images of floods, deep and widespread; floods of Biblical proportions. It comes from the Greek κατακλυσμος meaning a deluge. The *Oxford English Dictionary* states that, in geology, the term cataclysm means 'a great and general flood of water, a deluge', but that it is sometimes used in a rather vague way in the same sense as the term catastrophe, that is, as a sudden convulsion or alteration of physical conditions. Adjectives used in dictionaries to describe cataclysmic events include momentous, sudden, violent. In this book, a cataclysm will be used in the sense of a great and general flood of water occurring suddenly and violently. This usage of the term cataclysm conforms to the practice adopted by most writers on geological matters. However, one writer has recently suggested that it should be used in a different way: Benson (1984), in a noble attempt at sharpening the definitions of the words catastrophe and cataclysm, defines a catastrophe as an event which leads to a thoroughgoing reorganization of the components of an Earth system, without the system losing its identity; and he defines a cataclysm as an event which utterly destroys both the system and its components. The trouble with these definitions is that they are not etymologically correct. Consequently, they are at odds with the everyday usage of the words, and would probably create far more misunderstandings than they would resolve. For that reason, they will not be adopted here.

The words deluge, inundation, flood, and debacle are unambiguous enough, but it is worth spelling out their meanings so as to nip misunderstandings before they arise. (The *Oxford English Dictionary* is used as a source of definitions for these words, and where the definitions have been taken directly from that work, quotation marks are used.) The word deluge comes from the Latin *diluvium* via the French *déluge*. It means 'a great flood or overflowing of water, a destructive indundation'. It is sometimes used to denote the great Flood in the time of Noah, also called the general deluge or universal deluge. The word inundation comes from the Latin *inundationem* (*undare*, to flow), and means 'an overflow of water; a flood'. The term flood comes, through the Old Teutonic, from the Aryan verbal stem *plo* meaning the action of flowing. A flood is 'an overflowing or irruption of a great body of

water over land not usually submerged; an inundation, a deluge'.
It is also 'a profuse and violent outpouring of water; a swollen
stream, a torrent'. Finally, the word debacle, also spelt débâcle,
comes from the French *débâcler* meaning to unbar. It originally
signified the breaking up of ice on a river, but in geology nowadays
it is taken to mean 'any sudden flood or rush of water which breaks
down opposing barriers, and hurls forward and disperses blocks of
stone and other debris' (Page 1865: 169). A possible source of
confusion with the word debacle arises owing to its being used by
de Saussure and some other early geologists to mean, in effect, a
cataclysm (see p. 77); so long as the reader is alerted to this minor
departure of meaning, no problem should arise.

Perhaps the biggest problem with many of the words just
defined, is that they are used to denote floods and flooding in
general, and not just the biblical Flood. In an effort to make it
perfectly clear which flood is being alluded to, all mention of the
Noachian Flood, and other terms used to describe it, will be
signified by the use of a capital letter, as in universal Deluge,
general Flood, and so on. The only time where this practice may
fail, is in passages quoted from the works of other authors.

On methodology

Diluvialists are commonly equated with catastrophists. Although
there is some basis for this equation, it is misleading. Not all floods
are cataclysms. The flooding of the land may occur suddenly and
violently, *or* gradually and gently. To brand students of slow and
gentle floods with the same iron as students of catastrophic floods
seems unfair, and disguises important methodological and practical
differences between them. For this reason, it is worth elaborating
a little on the chief types of diluvialism.

Diluvialism is more than just the study of floods; it is a system of
Earth surface history. As such a system, it involves a number of
methodological presuppositions and substantive suppositions. In
any system of Earth surface history, one methodological presupposition holds. It is *the uniformity of law*, wherein natural laws
are held as invariant in space and time—the properties of energy
and matter are assumed to have been the same in the past as they
are at present. The second methodological presupposition involves

alternative assumptions. The first alternative, almost invariably chosen since about 1830, is the *the uniformity of process*, wherein, whenever possible, past effects are held as the outcome of causes seen in operation today; this is the assumption of actualism. It is also known as the principle of simplicity, which states that no additional properties should be postulated unnecessarily, and the principle of parsimony, or Occam's razor (Simpson 1970). The second alternative is a negative expression of the uniformity of process, and could be called the *non-uniformity of process*. It is the assumption taken by some of the catastrophic diluvialists who, in seeking a cause for the Noachian Flood, invoke the action of extraordinary, supernatural events which do not occur at present. Gould (1965) argues that both the uniformity of law and the uniformity of process are assumptions shared by all scientists, but this is not strictly true. Admittedly, both uniformitarians *and* catastrophists fervently supported the principle of uniformity of law (Rudwick 1972), but, while uniformitarians held staunchly to the principle of uniformity of process, catastrophists were equivocal about it, generally agreeing that present processes should be used to explain past events whenever possible, but, unlike the uniformitarians, being prepared to invoke, if necessary, causes which no longer operate.

Diluvialist systems of Earth surface history involve a substantive claim which, as in the claim concerning the uniformity of process, entails alternative assumptions. The claim concerns *the uniformity of rate* (Gould 1987: 120). It is assumed, either that the ordinary processes observed at present have acted at the same rate in the past—this is the assumption of gradualism; or else that presently observed processes have acted more intensely in the past, in a sudden and violent manner—this is the assumption of catastrophism. For example, gradualists would claim that large parts of continents could slowly be submerged, owing to a marine transgression. They would deny the catastrophists' claim that the land surface could suddenly be covered by water from a grand cataclysm. In fact— and this is an important point—both catastrophists *and* gradualists make definite suppositions about the rate of processes in the empirical world which may, or may not, be true.

There is actually another substantive claim, involving alternative assumptions, which is normally made in systems of Earth history. This is *the uniformity of state* (Gould 1987: 123). It is a claim concerning the direction of change: either the Earth has remained much the same throughout its history—this is assumption of non-directionalism, or dynamic steady-state; or else the Earth has changed in a definite direction—this is the assumption of directionalism. However, this particular claim is not a vital ingredient of diluvialist systems.

In summary, the kind of processes, and the rate of processes, acting at the Earth's surface, may all be assumed either to have been the same in the past as at present, or to have been different in the past. There are thus two fundamental dichotomies involved in systems of Earth surface history. They are:

The dichotomy of process: processes acting today may be assumed to have acted in the past, or may be assumed not to have acted in the past. In the latter case, different kinds of processes in the past must be invoked. The dichotomous assumptions about the uniformity of process give rise to either an actualistic or a non-actualistic methodology.

The dichotomy of rate: the intensity or rate of present processes may be assumed to have remained much the same throughout Earth history, or may be assumed to have varied. These dichotomous assumptions about uniformity of rate give rise to claims of either a constant (gradual, slow, steady) or a changing (catastrophic, paroxysmal) intensity of process.

To these dichotomies can be added a third, which, though not of great importance in diluvialist systems, is crucial in understanding the broader systems of Earth history of which diluvialism is a part. It is

The dichotomy of state: conditions at the Earth's surface today may be assumed to have remained much the same throughout Earth history, or they may be assumed to have changed. These dichotomous assumptions about the uniformity of state give rise to claims of either a constant (non-directional) or a changing (directional) state of the Earth's surface.

The least satisfactory of these dichotomies is the gradualistic–catastrophic split over the rate of process, since it represents two ends of a spectrum of possible levels of changes in rate—either slow and gentle change (what Gould calls stately change), or else abrupt and violent change. Nor does it allow for the possibility of gradual change having catastrophic results. None the less, it does seem to capture genuine differences in the systems of Earth surface history proposed by a wide range of early geologists, and in that context serves a useful purpose.

The double dichotomy admits of four possible diluvialist systems of Earth surface history. They are:

Non-actualistic, catastrophic diluvialism: this is the traditional diluvialism of most writers prior to about 1830, including Alexander Catcott and William Buckland. It assumes that, since the Creation, there has been just one, sudden, violent, and extraordinary event—the Noachian Flood—which sculptured most of the Earth's relief features, leaving sea shells on mountains, bones in caves and gravel deposits, polished and scratched rock surfaces, and depositing boulder clay and other diluvial sediments.

Non-actualistic, gradualistic diluvialism: this is a rare form of diluvialism in which the Flood, rather than being regarded as a cataclysm, is assumed to have been a gentle event, with the waters rising and retiring too slowly to remodel the Earth's topography. This view has been held by a few scholars, such as Nathanael Carpenter, relying on the evidence of Moses who describes the Flood as a quiet effusion of waters upon the face of the Earth.

Actualistic, catastrophic diluvialism: this is a system of diluvialism which was first proposed towards the end of the eighteenth century, when field observation revealed evidence of not one, but several cataclysms having occurred in Earth's history. Such a system was only possible when theological opinion had relaxed enough to permit far broader interpretations of the Scriptures. Proponents of this system, such as James Hall and Peter Simon Pallas, did not necessarily deny that a universal Deluge had occurred at the time of Noah, but they deemed it one of many such deluges, each of which had acted catastrophically. This is also the neodiluvialist system of Earth surface history, wherein cataclysms are seen as recurrent events, within the ordinary limits of nature, produced by meteorites crashing into the ocean.

Actualistic, gradualistic diluvialism: this is a system of diluvialism which was first adopted by Charles Lyell when he proposed his marine erosion theory. It is similar to actualistic, catastrophic diluvialism except that it rejects the claim that great floods act suddenly and violently, favouring instead the view that great floods occur slowly and gently. It is *not* the same as the system of Earth surface history which could be called uniformitarian diluvialism. The system of Earth history known as uniformitarianism is, in the strict usage proposed by James Hutton and Charles Lyell, an actualistic gradualism involving the assumption that the state of the Earth's surface has remained much the same throughout its entire history, though it does permit the repetition of geological conditions through successive epochs. It can be contrasted with evolutionism, a form of actualistic gradualism which, rather than focusing on the uniformity of the situation, stresses the uniformity of the change of the situation (Hooykaas 1970). Evolutionism is the system developed by Charles Darwin (1859), which enabled him to explain the non-uniform change of life throughout geological time in terms of almost uniform change. It was also the system followed by the geologists Bernhard Cotta (1846), and, in small measure, by Hutton, who expressed the view that Nature should not be limited by the uniformity of 'an equable progression' (Hutton 1788: 302).

On the defence of diluvialism

It would probably take a lot of searching to find the word diluvialism in any modern text on geology or geomorphology, except in books dealing with the historical development of those sciences. Even in the enlightened 1980s, most geologists and geomorphologists probably regard the terms diluvialism and catastrophism as historical curiosities which have little or no bearing on present-day research. Indeed, Alistair Pitty in his book *The nature of geomorphology* (1983) remarks that the term catastrophism, and by implication diluvialism, should be left in the nineteenth century with other Victorian paraphernalia. In the face of such views, it would seem necessary to explain why reviving diluvialism and delving into its past is deemed a worthwhile task.

There are at least two good reasons for studying the history of diluvialism. In the first place, to understand what goes on in Earth surface science today, it is necessary to appreciate the shifts of thought which have taken place since ideas concerning the history of the Earth's surface were first mooted. On this point, it is instructive to recall Isaac Newton's avowal that he could not have seen so far, had he not stood on the shoulders of giants. Archibald Geikie makes a similar point in a more elaborate manner:

In science, as in all other departments of enquiry, no thorough grasp of a subject can be gained, unless the history of its development is clearly appreciated. Nevertheless, students of Nature, while eagerly pressing forward in the search after her secrets, are apt to keep the eye too constantly fixed on the way that has to be travelled, and to lose sight and remembrance of the paths already trodden. It is eminently useful, however, if they will now and then pause in the race, in order to look backward over the ground that has been traversed, to mark the errors as well as the successes of the journey, to note the hindrances and helps which they and their predecessors have encountered, and to realise what have been the influences that have more especially tended to retard or quicken the progress of research. (Geikie 1905: 1)

In the second place, reopening the books of the diluvialists should prove salutary because they contain more substance than most modern writers suggest. Without doubt, ever since Lyell insinuated uniformitarianism into geology, the diluvialists have received unfair treatment, certainly in English-speaking nations. It is, of course, inevitable that all musings on the historical development of a discipline will be subjective, but there is evidence that the diluvialists have had a very raw deal from the blinkered uniformitarians. The views of the past masters of Earth surface history have been handed down to us in a very partial manner, and the passing references to the works of the diluvialists in modern textbooks, if reference is made at all, are usually scathing and erroneous. Even that worthy tome *The history of the study of landforms* (Chorley *et al.* 1964), splendid and entertaining though it is, leaves the reader with the impression that James Hutton, his advocate John Playfair, and Charles Lyell were the good guys who triumphed over the naughty catastrophic diluvialists, either by converting them to uniformitarianism, or by ignoring them. The history of geology has often been expounded, in the fashion of a fairy-tale, as a battle between good and evil (Hooykaas 1970). So has the history of geomorphology: catastrophic diluvialism is black; fluvialism and gradualistic diluvialism are white. This view wholly overlooks some of the very important points made by the catastrophic diluvialists, some of which, such as their explanations for the occurrence of quick-frozen mammoths in Arctic regions, even Lyell and Darwin found themselves at a loss to counter arguing from gradualistic principles (but see Kurtén 1986). It is very narrow-minded, as this book will demonstrate, to claim that fluvialism and gradualistic diluvialism are good, but catastrophic diluvialism is bad; that hypotheses couched in gradualistic terms are productive, whereas those couched in catastrophic terms are empty.

On the revival of diluvialism

Although a reconsideration of the views of the old diluvialists is of undoubted historical interest, there would be little point in attempting to revive the diluvialist cause (which many Earth scientists, certainly until the 1980s, had probably thought dead and buried long ago) unless, in a novel guise, it had something constructive to offer as a system of Earth surface history. A new and potentially productive brand of diluvialism has recently emerged, which may be called neodiluvialism (Huggett 1989). It has arisen indirectly from new developments in the field of planetary and space science. Space exploration has showed that bodies in the Solar System are heavily cratered, and space science has revealed that crater formation is still active on all the terrestrial planets and satellites, including the Earth and Moon. Before these discoveries, it was thought just possible that terrestrial catastrophes might be produced by a collision with a cosmic body. The astronomical discoveries have shown that bombardment does occur, and that it must have catastrophic consequences on the Earth. The result has been a revolutionary change in ideas concerning the Earth and its history. The bombardment hypothesis, from being the least likely explanation of catastrophes, has become, almost overnight, the ruling hypothesis; other possible hypotheses of terrestrial catastrophism, such as pole shift, being relegated to a very poor second place.

The bombardment hypothesis has many facets, only one of which is directly relevant to neodiluvialism—the possibility that meteorite impacts in the ocean may produce superwaves which rush over continental lowlands to create superfloods. The waters of these superfloods will fashion the landscape in the same way that the cataclysms so dear to the catastrophic diluvialists of old were thought to. Thus, the bombardment hypothesis casts a new light on some of the ideas of the old diluvialists, and provides the basis of a neodiluvialism (Huggett 1989).

In view of all the above arguments, it would appear not untimely to reassess diluvialism. The present book does not advertently provide any new and basic answers to problems in Earth surface history. It merely tries to show what diluvialism and neodiluvialism are about, and to offer the basis for alternative explanations for the history of the Earth's surface features. It does so by exploring the history of diluvialist beliefs from ancient chronicles, through the Renaissance, Restoration, Enlightenment, and eighteenth century, to the neodiluvialism of the late 1980s. Some readers may still question the wisdom of including old, outmoded ideas on diluvialism, and may think that it would be preferable to start the discussion in the twentieth century, or even in the 1980s. However, neodiluvialism represents the belated flowering of seeds sown long ago, and only by attempting to track the twists and turns in the development of diluvialist thought, can the implications of the neodiluvialism for studies of Earth surface history be evaluated.

On the chronology of diluvialist thought

To impart some order to the subject-matter of diluvialism, the views of the diluvialists will be treated in chronological order, and grouped into somewhat arbitrary periods. Diluvialist beliefs may be classed, like hymns, as ancient and modern. The ancient beliefs, which were mainly concerned with the acts of the Gods, will be discussed in Chapter 2. The modern era of diluvialist beliefs may be subdivided, like many of the geological periods, into early, middle, and late, with neodiluvialism, like the Holocene, being tagged on at the end.

The early modern period of diluvialism, which will be the subject of Chapters 3 and 4, starts in the Renaissance, though its roots tap some of the ideas discussed by medieval scientists, who themselves recovered the lost science of the ancient Greeks (see Adams 1938: 57–68), and progresses to the Restoration. During the early modern era, diluvialists thought that the hand of God was, directly or indirectly, responsible for the Flood, which was the only event, since the Creation, capable of making changes of any consequence to the Earth's face. Earthquakes, the collapse of subterranean caverns, the uplift of mountains, the shift of the poles, and the collision of the Earth with a comet were all considered as possible causes of the Flood. Thus, at this time, virtually all the possible agents of worldwide cataclysms were identified.

The middle modern period of diluvialism, which will be discussed in Chapters 4, 5, and 6, commences in the eighteenth century with the Age of Enlightenment, and involves the cosmogonies of the *philosophes* and other explanations of the biblical Flood. It then moves on to the first studies which professed to base their theories on evidence of the Noachian Flood, and other floods, found in the field: the studies of the French, English, and American schools of diluvialism; and the systems of Earth history espoused by the geognosists, in which universal floods are allotted an important role. The middle modern period of diluvialism had its heyday in the first quarter of the nineteenth century, but it came to a gradual end, in England and America at least, with the widespread acceptance of uniformitarianism during the middle decades of the nineteenth century.

Lyell's *Principles of geology* (1830–3) ushered in the late modern period of diluvialism. Explanations of floods on an Earth with a non-violent history, as proposed by the gradualistic diluvialists, will be examined in Chapter 7. Partial floods, floods produced by changes of climate, and debacles, will be discussed in Chapter 8. And finally, neodiluvialism, its nature, problems, and prospects, will be considered in Chapter 9.

2

The Seeds of Diluvialism
Floods in ancient writings

Flood myths

The Scottish geologist, Sir Archibald Geikie, in his work on *The founders of geology* (1905), explains that

The earliest efforts at the interpretation of nature found their expression in the mythologies and cosmogonies of primitive peoples, which varied in type from country to country, according to the climate and other physical conditions under which they had their birth. Geological speculation may thus be said to be traceable in the mental conceptions of the remotest pre-scientific ages. (Geikie 1905: 6)

Although the modern reader may demur at Geikie's environmental determinism, the tenor of his statement is unquestionably true. Ancient cultures in all parts of the world believed that Earth's history has included one or many catastrophes involving fire or flood, and often both. Cataclysms and conflagrations are recorded in many ancient mythologies, and are common in the traditions of almost all human races. It is perhaps not generally realized by Earth scientists just how universal flood myths are. One compilation lists over five hundred of them, belonging to over two hundred and fifty peoples or tribes. In most of these myths, there is a survivor of the flood, who, like Noah, is the progenitor of a new race of men. It would be pointless to describe all the flood myths, since a few samples will suffice to illustrate their ubiquity. Here are some which have been passed down in the traditions of 'primitive' tribes, described by Philip Freund in his book on *Myths of creation*:

Nichant, the Hero of the Gros Ventres, swims while holding onto a buffalo horn. Rock, the bold ancestor of the Arapaho Indians, fashions himself a boat of fungi and spiderwebs. The lone progenitor of the Annamese saves himself in a tom-tom. The Hero of the Ahoms in Burma uses a gigantic gourd which, by magical intervention, providentially grows out of a little seed. Trow, of the Tringus Dyaks of Borneo, is tossed on the waters in a trough; as is the Heroine of the Toradjas of Celebes, though hers is—most unromantically—a swill-trough. The ancestors of the Chané of Bolivia find refuge in an earthenware pot that floats. . . .

Other North American tribes as far apart as the Salinan and Chimariko Indians of California and the Crees of Manitoba and the Shawnees of Florida have similar stories. So do the Hurons, north of Lake Ontario, and the Algonkins along the St Lawrence. (Freund 1964: 10)

Three flood myths, recorded in ancient writings, are worth recounting in detail, chiefly because they may be a record of an actual cataclysm which occurred in Mesopotamia between 5000 and 4000 years ago (Rosenberg 1986: 21). In Europe and the Near East, the oldest flood myth appears to be *The epic of Gilgamesh*, a set of poems inscribed on clay tablets in the first centuries of the second millenium BC. The tablets, a veritable library of them, were unearthed at the Palace of Nineveh in 1839, by a young Englishman, Austen Henry Layard, and their cuneiform characters were eventually translated. A later American expedition, led by John Punnet Peters, excavating the ancient Nippur in southern Iraq, uncovered the oldest version of the Gilgamesh epic in the Sumerian language (Sandars 1960: 9–11). The story of the great flood is told to Gilgamesh by the flood survivor, Utnapishtim. It relates how the gods, aroused by the clamour of the multitude of people in the world, agreed to exterminate mankind, and how Enlil, the warrior god, did so. But Utnapishtim had been forewarned of the forthcoming destruction. In a dream, the god Ea had told him to tear down his house of reeds and build a boat. Ea had given precise instructions as to the size and shape of the boat, and had said that it must be filled with the seed of all living creatures. The boat was built according to the divine blueprint, and loaded with Utnapishtim's family, kin, all wild and tame beasts, and craftsmen. In the evening of the appointed day, Utnapishtim tells Gilgamesh,

the rider of the storm sent down the rain. I looked out at the weather and it was terrible, so I too boarded the boat and battened her down. All was now complete, the battening and the caulking; so I handed the tiller to Puzur-Amurri the steersman, with the navigation and the care of the whole boat.

With the first light of dawn a black cloud came from the horizon; it thundered within where Adad, lord of the storm was riding. In front over hill and plain Shullat and Hanish, heralds of the storm, led on. Then the gods of the abyss rose up; Nergal pulled out the dams of the nether waters, Ninurta the warlord threw down the dykes, and the seven judges of hell, the Annunaki, raised their torches, lighting the land with their livid flame. A stupor of despair went up to heaven when the god of the storm turned daylight to darkness, when he smashed the land like a cup. One whole day the tempest raged, gathering fury as it went, it poured over the people like the tides of battle; a man could not see his brother nor the people be seen from heaven. Even the gods were terrified at the flood, they fled to the highest heaven, the firmament of Anu; they crouched against the walls, cowering like curs. . . .

For six days and six nights the winds blew, torrent and tempest and flood overwhelmed the world, tempest and flood raged together like warring hosts. When the seventh day dawned the storm from the south subsided, the sea grew calm, the flood was stilled; I looked at the face of the world and there was silence, all mankind was turned to clay. The surface of the sea stretched as flat as a roof-top; I opened a hatch and the light fell on my face. Then I bowed low, I sat down and I wept, the tears streamed down my face, for on every side was the waste of water. I looked for land in vain, but fourteen leagues distant there appeared a mountain, and there the boat grounded; on the mountain of Nisir the boat held fast, she held fast and did not budge. One day she held, and a second day on the mountain of Nisir she held fast and did not budge. . . . When the seventh day dawned I loosed a dove and let her go. She flew away, but finding no resting-place she returned. Then I loosed a swallow, and she flew away but finding no resting-place she returned. I loosed a raven, she saw that the waters had retreated, she ate, she flew around, she cawed, and she did not come back. Then I threw everything open to the four winds, I made a sacrifice and poured out a libation on the mountain top. (Translated in Sandars 1960: 110–11)

The Greeks, too, had a flood myth, told in the flood cycle (Rosenberg 1986; 22–7). Zeus was angered when, on travelling among a race of men which he had heard cared only about themselves, he was insulted by King Lycaon who lived in Arcadia. Zeus returned to Mount Olympus, bent on punishing all mortals for their evil ways. He commanded Aeolus to restrain the fair-weather winds, and to release the storm winds to destroy all crops. This having been done, Zeus was still enraged, and he asked Poseidon, his brother, to let free the rivers and streams to spill over all barriers and rush unrestrained upon the land. Higher and higher the flood waters rose, till the Earth was a vast sea. All mortals drowned, save Deucalion and his wife Pyrrha. Deucalion had been forewarned of the coming flood by his father Prometheus, whose fate it was to see all future events except those pertaining to his own immortal life. Prometheus counselled Deucalion to build a very large chest and to stock it with provisions of all kinds. When the flood came, Deucalion and Pyrrha boarded the chest. After nine days, the waters, by Zeus's commands, were made calm and ordered to retreat. Deucalion and Pyrrha sighted the twin peaks of Mount Parnassus, which had escaped from the flood waters because they reached through the clouds, and moved their floating chest towards them. They came to rest on the side of the mountain, and stood on firm land, the only survivors of Deucalion's flood.

The best-known flood myth is probably the one described in Genesis. As this story was such a source of inspiration to the diluvialists, it is worth repeating it here, starting at the point where Noah has, according to God's instructions, made the ark:

And it came to pass after seven days, that the waters of the flood were upon the earth. In the six hundredth year of Noah's life, in the second month, the seventeenth day of the month, the same day were all the fountains of the great deep broken up, and the windows of heaven were opened. And the rain was upon the earth forty days and forty nights. In the selfsame day entered Noah, and Shem, and Ham, and Japheth, the sons of Noah, and Noah's wife, and the three wives of his sons with them, into the ark; they, and every beast after his kind, and all the cattle after their kind, and every creeping thing that creepeth upon the earth after his kind, and every fowl after his kind, every bird of every sort. And they went in unto Noah into the ark, two and two of all flesh, wherein is the breath of life. And they that went in, went in male and female of all flesh, as God had commanded him: and the LORD shut him in. And the flood was forty days upon the earth; and the waters increased, and bare up the ark, and it was lift up above the earth. And the waters prevailed, and were increased greatly upon the earth; and the ark went upon the face of the waters. And the waters prevailed exceedingly upon the earth; and all the high hills, that were under the whole heaven, were covered. Fifteen cubits upward did the waters prevail; and the mountains were covered. . . . And the waters prevailed upon the earth an hundred and fifty days. (Genesis 7: 10–24.)

And God remembered Noah, and every living thing, and all the cattle that was with him in the ark: and God made a wind to pass over the earth, and the waters assuaged; The fountains also of the deep and the windows of heaven were stopped, and the rain from heaven was restrained; And the waters returned from off the earth continually: and after the end of the hundred and fifty days the waters were abated. And the ark rested in the seventh month, on the seventeenth day of the month, upon the mountains of Ararat. And the waters decreased continually until the tenth month: in the tenth month, on the first day of the month, were the tops of the mountains seen. And it came to pass at the end of forty days, that Noah opened the window of the ark which he had made: And he sent forth a raven, which went forth to and fro, until the waters were dried up from off the earth. Also he sent forth a dove from him, to see if the waters were abated from off the face of the ground; but the dove found no rest for the sole of her foot, and she returned unto him into the ark, for the waters were on the face of the whole earth: then he put forth his hand, and took her, and pulled her in unto him into the ark. And he stayed yet another seven days; and again he sent forth the dove out of the ark; and the dove came into him in the evening; and, lo, in her mouth was an olive leaf plucked off: so Noah knew that the waters were abated from off the earth. And he stayed yet another seven days; and sent forth the dove; which returned not again unto him any more. And it came to pass in the six hundredth and first year, in the first month, the first day of the month, the waters were dried up from off the earth: and Noah removed the covering of the ark, and looked, and, behold, the face of the ground was dry. And in the second month, on the seven and twentieth day of the month, was the earth dried. (Genesis 8: 1–14)

The belief in Noah's Flood and Deucalion's flood influenced, in a greater or lesser degree, 'the speculations of the philosophers who began to observe the operation of natural processes and who, though their deductions were often about as unscientific as the myths for which they were substituted, may yet be claimed as the earliest pioneers of geology' (Geikie 1905: 6–7). Ideas scattered through the literature of Greece and Rome reveal the first stages of advance in theoretical opinions concerning fire myths and flood myths. For instance, Anaximander (610–540 BC), Anaximenes (570–? BC), and his pupil Diogenes of Apollonia (lived in fifth century BC), all members of the Ionian school of Greek philosophy, believed that the world suffers destruction and subsequent re-creation. Heraclitus (540–475 BC), also of the Ionian school, taught that the world is destroyed by fire every 10 800 years. Plato (429 or 427–347 BC) relates in his *Timaeus* the theory that there have been, and will be again, periodic annihilations of the earth by fire and flood; he attributes these catastrophes to the action of a celestial body passing near the Earth. The source of this theory is Solon of Athens who, on a visit to Egypt, questioned priests about

Earth history. One of the priests explained that there have been and there will be many and varied destructions of mankind, of which the greatest are by fire and water. He pointed out that the story of Phaeton, though regarded as a legend, has a grain of truth which lies in the occurrence of a shifting of the bodies in the heavens which move around the Earth, and a destruction of the things on the Earth, which recurs at long intervals. Phaeton, the son of Helios and Prote, obtained his father's permission to drive his father's chariot (the Sun) across the heavens, but, being unable to check the horses, burnt up all that was on the Earth. He was slain by Zeus with a thunderball for his efforts.

The meaning of myths

It is exceedingly difficult to say just why so many ancient cultures should have believed in cataclysms. Myths in general have been interpreted in several different ways. Traditionally, they were regarded as divine revelations, and indeed still are in many religious circles. The philosophers of the eighteenth century deemed them barbaric superstitions and sought to sweep them under the carpet, but they were thwarted in this endeavour by the members of the Romantic movement who rediscovered the beauty of the myth. And in the late nineteenth and early twentieth centuries, scholars 'began to develop some daring new ideas about myth' (Freund 1964: 16). These radical views about myths fall into three basic categories: myth as art and philosophy, as examples of early man's creative imagination at work, as exciting pieces of fiction; myth as primitive science, as early man's ingenious efforts to account for the world around him; and myth as literal history, as factual accounts of events which occurred during man's early tenure of the Earth. Of course, the question of the meaning of myths is far more complicated than these simple categorizations would suggest, involving also myth as ritual, as a manifestation of inhibitions, and as socially pragmatic tales; but further exploration of such matters is beyond the scope of the present discussion. It suffices merely to report that, if the ancient myths are taken at face value, then the view that the Earth has undergone several cataclysmic events during its history was common among ancient cultures.

The myths have been extensively picked over by post-Renaissance scholars for information which may shed light on the history of the Earth. Many of the old catastrophists drew heavily on the Scriptures, and other early writings, as a fertile source of information concerning the history of the Earth. Robert Hooke, William Whiston, John Woodward, Alexander Catcott, and almost all the great cosmogonists of the seventeenth and eighteenth centuries make frequent references to Genesis and the classical literature. As more and more of the world was explored, so new traditions and myths were collected. For example, in a book entitled *L'Antiquité dévoilée par ses usages, ou examen critique des principales opinions, cérémonies et institutions religieuses et politiques des différens peuples de la terre*, published in 1766, Nicolas-Antoine Boulanger analysed the cosmogonies and mythologies of many far-spread cultures, including Germans, Greeks, Jews, Arabs, Hindus, Chinese, Japanese, Peruvians, Mexicans, and Caribs. He concluded from these stories that the human race had been subject to a series of cosmic convulsions, supporting his claim with evidence from the geological and palaeontological record of Earth history. So, during the eighteenth century, the writings of an increasing range of cultures were quoted in works dealing with the history of the Earth. Even Charles Lyell felt pleased to describe some of the ancient systems of the world, in the first volume of his *Principles of geology* (1830).

Evidence of floods in the classical landscape

The younger geological formations in the countries adjacent to the Mediterranean Sea underlie the lowlands and outcrop high along the flanks of hills and mountains. These formations, having been upraised from beneath the sea, are teeming with the fossil remains of shells and other marine creatures. Many of these fossils closely resemble organisms still living in neighbouring seas. It is hardly surprising, therefore, that the early Greek philosophers, after examining these fossiliferous deposits, concluded that the sea had once covered the tracts of land where the marine shells are found. Indeed, there can be little doubt that the presence of marine shells high in hills and mountains 'led to those wide views of the vicissitude of Nature' which were adopted by the successors to the Greeks in later centuries (Geikie 1905: 11).

The first records and interpretations of marine fossils survive in the fragmentary remnants of the writings of the early Greek 'Sophi' or 'Wise Men', later to be called philosophers. These men lived during the 'salad years' of Greek philosophy, from 585 BC until about 400 BC; they were 'green and genial individuals' who established the scope and determined the problems of philosophy (Barnes 1987: 9). The literature of these early Greeks was undoubtedly extensive, but a large part of it has been lost. The views of many famous names survive as hearsay, as tantalizing extracts reported by later classical commentators. The views of many others survive as threads and patches, as Adams (1938: 9) puts it. From these fragments of original writings and later reports can be gleaned a very crude idea of the teachings of the early Greeks. Xenophanes, from Colophon in Ionia, who lived around 540–510 BC, is reported by Hippolytus in his *The Refutation of all heresies* (1870 edn.: 1. 12) as having written concerning sea shells found in the middle of the land and on mountains. He mentioned impressions of fish and seaweed and the remains of seals found in the quarries of Syracuse, an impression of a bay-leaf found deep in the rocks of Paros, and the shapes of all sea creatures on Malta. From these observations, he concluded that the land is periodically submerged, and then earth is carried down to the sea and becomes mud. It is in this mud that the impressions are made. Xanthus, from Sardis in Lydia, who lived around 480 BC, is quoted by Strabo (64 BC–AD 25?), the famous Greek geographer and historian from Amasia in Asia Minor who worked in Rome, in his *Geography* (1966–70 edn.: 1. 3. 4), as having seen shells like cockles and scallops, far from the sea, in Armenia, Phrygia, and Lydia; and having inferred from this evidence, and that of scattered salt lakes, that these regions had once been submerged beneath the sea. Herodotus, an approximate contemporary of Xanthus, noted abundant nummulites in the Eocene limestone around the pyramids of Egypt, especially near the oasis of Jupiter Ammon. He regarded these fossils as the indurated remains of lentils which had been stored to provide food for the builders of the pyramids. From the presence of the 'indurated remains of lentils', and from the presence of a saline crust on the ground, he concluded that the sea had once spread over lower Egypt. Strabo, in his *Geography*

(1966–70 edn.: 1. 3. 4), reported Eratosthenes' (276–190 BC) confirmation of Herodotus' observations. Eratosthenes found vast quantities of marine shells on the road to the Ammon oasis, together with beds of salt and saline springs. Strabo himself was widely travelled and concluded from his extensive observation that in certain areas the land has subsided and in others it has been elevated. To this elevation he attributed the presence of marine shells in rocks which are now well above the level of the sea. Writing of the periodic inundation of large regions of the sea, he explains that all things are in perpetual motion and suffer great changes, much of the sea being turned into land, much of the land being turned into sea. Similar opinions were expressed in the fifteenth book of Ovid's *Metamorphoses*, in which the poet speaks for Pythagoras, the very early Greek philosopher who failed to set down his ideas in writing. Pythagoras is reported to have held that the elevation and depression of great areas of land has led to former parts of the mainland becoming covered by the sea, and former seas becoming dry land.

The best-documented views on the interchange of land and sea were expressed by the great naturalist and philosopher, Aristotle. Aristotle (384–322 BC) belonged to the second period of Greek philosophers, the period of the Schools, which ended about 100 BC (Barnes 1987: 9). In his *Meteorologica* (1931 edn.: 1. 14), Aristotle explains that the sea now covers areas which were once dry land, and that land will one day reappear where today sea is found. He sees an order and periodicity to these transpositions of land and sea, likening the process to life histories of animals and plants with their periods of vigour and decline, but pointing out that, whereas an entire organism grows and then dies, the surface of the Earth is affected only locally. He adds that the interchanges of land and sea take place too slowly, and over far too protracted a time, to be noticed during a man's brief lifetime. As to the cause of the terrestrial mutations, Aristotle writes that, just as winter regularly recurs among the march of the seasons, so a great winter, lasting through an enormous period of time, may arise, bringing with it very heavy rainfall. The excessive rains do not affect all countries, but where they do fall, devastating floods, such as that survived by Deucalion in the area of old Hellas, will occur. Aristotle styles this great winter the *kataklysmos*, meaning the deluge. He also refers to a great summer, or *ekpyrosis* (conflagration), when, the lowlands of some regions, mainly owing to the porous nature of the rocks, become more or less desiccated and stay so until the great winter returns.

The Flood in the Dark Ages

The ancient and modern eras of diluvialist thought are interrupted by the Dark Ages. This hiatus in speculations and studies of floods lasted from the fall of the Roman Empire to the revival of learning in fifteenth-century Europe. During this period, knowledge survived in monasteries and other ecclesiastical establishments. But, with the exception of the natural history found in the works of the Greek and Roman scholars, knowledge of nature was of little concern. The feeble flame of interest in the study of nature was rekindled and kept alight by the Arabs from the middle of the eighth century onwards for some five hundred years: with great labour and at large expense, the Arabs procured as much of the literature of Ancient Greece and Rome as they could, and translated into Arabic the works of the best classical philosophers, physicians, mathematicians, and astronomers (cf. Geikie 1905: 42). The most illustrious Arab scholar was Avicenna, also known as Ibn-Sina (980–1037), who translated Aristotle. It is important to realize that, during the Middle Ages in Europe and in the Arabian states, Aristotle was regarded as the chief among philosophers, and one whose opinions on any subject were authoritative and final (Adams 1938: 16). Only with the Renaissance did scholars start to question the views of the classical writers. It was during the Renaissance that modern diluvialist ideas were first expressed, and it to this early rise of diluvialism that discussion will now turn.

3

The Rise of Diluvialism
Floods and Renaissance scholars

The background to Renaissance 'geology'

Theories of the Earth proposed by cosmogonists belonging to the period of the Renaissance and up to the Restoration are sketchy and incomplete. Seldom are they set down clearly. Rather, they are contained in treatises on other topics, or are given a passing mention in philosophical digressions. Modern readers find them quaintly artless. As Frank Dawson Adams quips in his tome on *The birth and development of the geological sciences*,

These early fables of geological science should be read by all who are in need of mental recreation and who possess the required leisure and a certain sense of humor, although many of them make a further demand upon the seeker after amusement and recreation in this fairyland of science, namely, that he shall seek his relaxation in the somewhat unaccustomed field of medieval Latin. (Adams 1938: 210)

Lyell is no less uncharitable about early theories of the Earth. Although he has a high regard for the conduct of scholarly disputes in the sixteenth century, he deplores the manner in which they were argued:

The system of scholastic disputations encouraged in the Universities of the middle ages had unfortunately trained men to habits of indefinite argumentation, and they often preferred absurd and extravagant propositions, because greater skill was required to maintain them; the end and object of such intellectual combats being victory and not truth. No theory could be so far-fetched or fantastical as not to attract some followers, provided it fell in with popular notions; and as cosmogonists were not at all restricted, in building their systems, to the agency of known causes, . . . [they could feign] imaginary causes, which differed from each other rather in name than in substance. (Lyell 1834: i. 35–6)

Although it is impossible not to read fifteenth- and sixteenth-century writings without the distortion caused by nineteenth- and twentieth-century spectacles, it is grossly unfair to ridicule them. In their own time they were an integral part of the complex phenomenon known as the Scientific Revolution, and it is in that context that they should be evaluated (Kelly 1969: 215). It is certainly necessary to bear in mind that, during the fifteenth and sixteenth centuries, all knowledge concerning the causes of natural phenomena was generally supposed to have been given to man by God, and the message was closely guarded by the Church. Three important facts, accepted by all but the heretical, were laid down in the Scriptures: the Earth was a mere few thousand years old; since the Creation, the Noachian Flood was the only great catastrophe capable of wreaking considerable changes of the Earth's surface; and the world would be destroyed by a conflagration at the end of the millenium. Most early writers who refer to the origin and final destruction of the Earth accept without question the events described in Genesis. Examples of this blind acceptance of the Scriptures is evident in the writings of such theological giants as Martin Luther (1483–1546), for instance in *Luther's commentary on Genesis* (Luther 1958 edn.), and John Calvin (1509–64), for example in his *Commentaries on the First Book of Moses called Genesis* (Calvin 1948 edn.).

It also necessary, when evaluating the early views about the Earth, to bear in mind that the system proposed by Empedocles and elaborated by Aristotle, which maintained that there were four elements—air, earth, fire, water—was still accepted, and would be until the seventeenth century. Thus, to

most men before, during, and for some time after the Renaissance, the earth was a solid, spherical body composed of assorted metals, rocks, and earth (all of which were predominantly formed of the element earth, with varying proportions of the other elements). Within this solid body were underground regions of water, air, and fire. Bodies of water covered some areas of the earth, a sphere of air surrounded the earth, and a sphere of fire surrounded the air. (Kelly 1969: 217)

A few writers did question the accepted system of the world. Some did not accept Aristotelian cosmology; some did not think that Genesis should be interpreted literally; and some accepted neither Aristotle nor Genesis (Kelly 1969: 216). Nicolas of Cusa (1401–64), for instance, saw creation as one continuous act of God (Cusanus 1954), while Giordano Bruno (1548–1600) envisaged a world consisting of monads, basic units forever forming, moving, and reforming (Singer 1950). Other writers sought to account for the events described in the Scriptures using natural causes. Notable among these writers are Philippus Aureolus Paracelsus (1493–1541), originally named Theophrastus Bombastus von Hohenheim, and Nathanael Carpenter (1589–1628?). Paracelsus, the celebrated Swiss-German physicist and alchemist, who reputedly kept a familiar or small demon in the hilt of his sword, accepted that the world had been created in an instant by God, but believed that the water was made first, and then from it the heavens and Earth (Paracelsus 1951 edn.). Nathanael Carpenter, in his *Geography delineated forth in two bookes: Containing the sphaericall and tropicall parts thereof* (1625), a treatise important only in that it is acclaimed as the first pioneer British work on theoretical aspects of geography (Davies 1969: 29), conjectured that when God first created the Earth, its surface was round, uniform, and all covered with water. God then made the land into a 'pleasing shape' by fashioning cavities and mountains in the smooth, solid core.

The theories of Carpenter and Paracelsus are not of signal importance, but the fact that they were allowed to be promulgated reflects the mild liberalism which, by the sixteenth century, laid the Scriptures open to a slightly less strict interpretation. Another sign of this less rigid attitude towards the Scriptures was the postponing of the date of the final conflagration:

we find, in the speculations of early geologists, perpetual allusion to such an approaching catastrophe; while in all that regarded the antiquity of the earth, no modification whatever of the opinions of the dark ages had been effected. Considerable alarm was at first excited when the attempt was made to invalidate, by physical proofs, an article of faith so generally received; but there was sufficient spirit of toleration and candour amongst the Italian ecclesiastics, to allow the subject to be canvassed with much freedom. (Lyell 1834: i. 35)

None the less, the increased latitude given to the interpretation of the Scriptures was not without bounds. Giordano Bruno denied the existence of the Noachian Flood, and for this heretical act he burned at the stake (Winter 1916: 169). Outright contradiction of biblical teachings was thus not tolerated. But careful reinterpretation was permissible, though even then the exercise of caution was advisable. Winter (1916: 169) thinks it unlikely that Nicolaus

Steno would have escaped persecution for the views expressed in his *Prodomus*, despite his taking pains to make his views compatible with the Scriptures, had he not been a devout Catholic and under the protection of a powerful grand duke.

Few writers of the Renaissance–Restoration period chose to deny that the Noachian Flood had occurred, which under the circumstances was a rather wise move. However, the nature of the Noachian Flood, and its effect upon the Earth's surface features, were discussed at length. It was during this early period of speculation over the nature of the Noachian Deluge that the seeds of diluvialist thinking were planted, seeds which were to bear fruit in the nineteenth century. It is these very early views on diluvialism which will now be considered.

The nature of the Flood

In seventeenth-century Britain, the effects that the Flood had had on the Earth's surface were the subject of a lively discussion. Two schools of thought emerged. One school, which included among its members the Welsh topographer George Owen of Henllys (1552–1613), believed that the Flood was 'a divine instrument which in a single climacteric act had destroyed not only all pre-diluvial life, but also the pre-diluvial world itself' (Davies 1969: 38). The most vehement advocates of diluvial metamorphosis, as Davies (1969) calls it, pictured the Flood as

a vicious, swirling body of waters sweeping over the globe, tearing up the smooth surface of exquisite pristine continents, twisting and shattering the Earth's rocks, and finally leaving the debris to form the world's present topography. (Davies 1969: 39)

The other school, which included among its number the geographer Nathanael Carpenter, opined that the waters of the Flood were calm and incapable of refashioning the Earth's topography, but they would, as Noah learns, 'destroy all flesh, wherein is the breath of life, from under heaven' (Genesis 6: 17). Carpenter (1625) allowed that not all the present rivers and fountains had existed from the beginning of the world, but would not countenance major changes of the Earth's topography. He believed that the Flood had had little effect in the sculpturing of the Earth's surface. The basis of his belief was that if the Earth's surface had been radically altered by the Flood, then the antediluvian mountains and rivers would not be the same as the postdiluvian ones, and so would bear different names. As they do not have different names, the Flood cannot have caused them to change greatly. Carpenter also argued that the waters of the Flood were not strong enough to move large quantities of earth. As evidence of the placidity and ineffectual power of Flood waters in scouring the Earth, he pointed out that the olive-leaf brought to Noah by a dove, and taken as a sign that the Flood waters were retiring, must have come from a tree still rooted in the ground: the Flood, it would seem, was unable to uproot an olive tree, so how could it possibly have thrown up lofty mountains or made land out of the former floor of seas? This somewhat suspect evidence for the placidity of the Flood was used as late as the early nineteenth century by Lyell in his *Principles of Geology* (1830–3).

Evidence of the Flood

Fossil sea shells on Italian mountains

In the early sixteenth century, it was generally believed that marine fossils found in mountains and plains were the product of the Noachian Flood. At about this time, a number of European, and particularly Italian, thinkers began to show an interest in geological phenomena, especially fossils. The full history of the study of fossil shells in Italy, from 1300 to 1810, is described by Giovanni Battista Brocchi (1772–1826), in the two volumes of his *Conchyliologia fossile subapennia*, published in Milan in 1814. The very brief outline given below is based chiefly on the historical surveys written by Lyell (1834: vol. i), Geikie (1905), and Gortani (1963).

Arguments over the true nature of the petrified Italian sea shells are confounded with arguments over the part played by the Noachian Flood in Earth history. It was generally held by early Italian scholars that the shells found on mountains were carried there by the Noachian Deluge. Most scholars regarded the Flood as a supernatural event, the cause of which need not be questioned any further. A few bolder thinkers suggested ways in which

cataclysms might be produced naturally. For instance, the Neapolitan jurist and antiquarian, Alessandro degli Alessandri (1661–
1523), in his *Dies geniales* (1522), noted that there were fossil shells
in the Calabrian mountains, and ascribed their presence there to
flooding by the ocean. He attributed the flood to a catastrophe, or
to a geographical change in the axis of rotation of the Earth (von
Zittel 1901: 14). But such views were rare, because, as has been
explained, it was almost unthinkable during the Renaissance to
question the Scriptures or even to interpret them in a new light,
such was the power of the Church in totally squashing opinions
which appeared to run counter to orthodox beliefs:

If therefore an observer who found abundant sea-shells imbedded in the
rocks forming the heart of a mountain chain ventured to promulgate his
conclusion that these fossils prove the mountains to consist of materials
that were accumulated under the sea, after living creatures appeared on
the earth, he ran imminent risk of prosecution for heresy, inasmuch as
according to Holy Writ, land and sea were separated on the third day of
creation, but animal life did not begin until the fifth day. Again, the
overwhelming force of the evidence from organic remains that the
fossiliferous rocks must have taken a long period of time for their
accumulation could not fail to impress the minds of those who studied the
subject. But to teach that the world must be many thousands of years old
was plainly to contradict the received interpretation of Scripture that not
more than 6000 years had elapsed since the time of the Creation.
    To court martyrdom on behalf of such speculative opinions was not a
course likely to be followed by many enthusiasts. Various shifts were
accordingly adopted, doubtless in most cases honestly enough, in order to
harmonize the facts of Nature with what was supposed to be the divine
truth revealed in the Bible. (Geikie 1905: 44–5)

Pious though many early geologists might have been, they must
surely have found it difficult to reconcile the Scriptures with the
evidence they unearthed in the rocks. To be sure,

There were many observers . . . who could not gainsay the evidence of
their own senses, and who recognized that either we must believe that the
minute and perfect-preserved organic structures in the fossils could only
have belonged to once living plants and animals, like those which possess
similar structures at the present day, or that the Creator had filled the
rocks of the earth's crust with these exquisitely designed but deceptive
pieces of mineral matter, with no apparent object unless to puzzle and
disconcert the mind of frail humanity. (Geikie 1905: 45–6)

Thus, the Renaissance geologists were faced with the problem of taking fossils as the remains of former plants and animals, and so coming into conflict with the Church; or accepting that they were created by God, or Satan, to deceive the unwary, and so going against their geological instincts as to the true nature of fossils. An obvious solution to this dilemma presented itself to those early orthodox minds: invoke the Noachian Flood to explain the distribution of marine fossils on mountains. At that time, the Flood was regarded as a worldwide cataclysm. For those writers who

had little or no personal acquaintance with the actual conditions of the problem, who did not realize the orderly manner in which the fossils are disposed, layer upon layer, for thicknesses of many thousand feet in the solid rocks of the land, the doctrine of the efficiency of the Flood offered a welcome solution of the difficulty. They had no conception of the physical impossibility of accumulating all the fossiliferous formations of the earth's crust within the space of one hundred and fifty days . . . It was enough for them to obtain warrant from Scripture that, since the creation of animal life, the dry land had been submerged, and to adduce evidence from the rocks which they could claim as striking corroboration of the truth of the biblical story. (Geikie 1905: 46).

Thus arose the first true diluvialists—students of the Flood—who contended that the Noachian Cataclysm was a potent geological event in the history of the Earth, indeed, the only event of any significance since the Creation. As shall be seen, the influence of the diluvialists on the development of geology has been profound. Their ideas proved popular, perhaps more because they seemed to offer a system of Earth history which actively supported the Scriptures, than because they were an exceptionally convincing explanation of the field evidence. Most historians of geology argue that the advance of rational concepts to explain the fundamental facts of historical geology was hampered by the diluvialists and the fruitless arguments they stirred up over the nature of organic remains and the role of the Flood in fashioning the Earth's surface and distributing marine fossils over mountains. Be that as it may, there is reason now to believe that cataclysms might occur, and although the early diluvialists might have been arguing their case from an unscientific premiss—the Scriptures state that the Flood occurred, therefore it did occur—some of the evidence they adduced in support of their claims might well be the product of recent cataclysms. This contentious issue will be discussed in Chapter 9.

A non-diluvial origin of Italian sea shells

There were some scholars who, having seen the great extent of the stratigraphical succession, found it impossible to follow the diluvialists in believing that such massive accumulations of rocks, teeming with fossils, could be produced by such a transient event as the Noachian Flood. The first person to question the diluvial origin of sea shells found in mountains was, probably, the illustrious painter, architect, sculptor, and engineer, Leonardo da Vinci (1452–1519). In his youth, Leonardo had been employed as an engineer in the construction of canals in northern Italy, and had seen numerous fossils embedded in the rocks into which the canals were cut. Thus his mind was alerted to the problem concerning the origin of marine fossils. He was unconvinced that the fossils were of diluvial origin. Rather, he favoured the view that changes in the level of land and seas had led to the inland and upland distribution of marine fossils, an idea which, as was mentioned in the previous chapter, had first been mooted by the early Greek philosophers, Aristotle in particular. Leonardo's arguments are very advanced for their time and are worth reporting in detail:

If you should say that the shells which are visible at the present time within the borders of Italy, far away from the sea and at great heights, are due to the Flood having deposited them there, I reply that, granting this Flood to have risen seven cubits above the highest mountains, as he has written who measured it, these shells which always inhabit near the shores of the sea ought to be found lying on the mountain sides, and not at so short a distance above their bases, and all at the same level, layer upon layer.

Should you say that the nature of these shells is to keep near the edge of the sea, and that as the sea rose in height the shells left their former place and followed the rising waters up to their highest level:—to this I reply that the cockle is a creature incapable of more rapid movement than the snail out of water, or is even somewhat slower, since it does not swim, but makes a furrow in the sand, and supporting itself by means of the sides of this furrow it will travel between three and four braccia in a day; and therefore with such a motion as this is it could not have travelled from the Adriatic sea as far as Monferrato in Lombardy, a distance of two hundred and fifty miles in forty days,—as he has said who kept a record of that time.

And if you say that the waves carried them there,—they could not move by reason of their weight except upon their base. And if you do not grant me this, at any rate allow that they must have remained on the tops of the highest mountains, and in the lakes which are shut in among the mountains, such as the lake of Lario or Como, and Lake Maggiore, and that of Fiesole and of Perugia and others. . . .

If you should say that the shells were empty and dead when carried by the waves, I reply that where the dead ones went the living were not far distant, and in these mountains are found all living ones, for they are known by the shells being in pairs and by their being in a row without any dead, and a little higher up is the place where all the dead with their shells separated have been cast up by the waves. Near to there the rivers plunged into the sea in great depth; like the Arno which fell from Gonfolina near to Monte Lupo and there left gravel deposits, and these are still to be seen welded together, forming of various kinds of stones from different localities and of varying colour and hardness one concrete mass. And a little farther on, where the river turns towards Castel Fiorentino the hardening of the sand has formed tufa stone; and below this it has deposited the mud in which the shells lived; and this has risen in layers according as the floods of the turbid Arno were poured into this sea, and so from time to time the bed of the sea was raised.

This has caused these shells to be produced in pairs, as is shown in the cutting of Colle Gonzoli, made sheer by the river Arno which wears away its base, for in this cutting the aforesaid layers of shells may be seen distinctly in the bluish clay, and there may be found various things from the sea. (da Vinci 1977: i. 314–15)

Da Vinci is specific about the inability of the Flood to have transported the shells:

And the Flood could not have carried them there because things heavier than water do not float upon the surface of the water, and the aforesaid things could not be at such heights unless they had been carried there floating on the waves, and that is impossible on account of their weight.

Where the valleys have never been covered by the salt waters of the sea there the shells are never found; as is plainly visible in the great valley of the Arno above Gonfolina, a rock which was once united with Monte Albano in the form of a very high bank. This kept the river dammed up in such a way that before it could empty itself into the sea which was afterwards at the foot of this rock it formed two large lakes, the first of which is where we now see the city of Florence flourish together with Prato and Pistoia; and Monte Albano followed the rest of the bank down to where now stands Serravalle. In the upper part of the Val d'Arno as far as Arezzo a second lake was formed and this emptied its waters into the above-mentioned lake. It was shut in at about where we now see Girone, and it filled all the valley above a distance of forty miles. This valley received upon its base all the earth carried down by the turbid waters and is still to be seen at its maximum height at the foot of Prato Magno for there the rivers have not worn it away.

Within this soil may be seen deep cuttings of the rivers which have passed there in their descent from the great mountain of Prato Magno; in which cuttings there are no traces visible of any shells or of marine earth. (da Vinci 1977: i. 316–17)

The controversy over the diluvial origin of the Italian fossil sea shells became animated in 1517. Fossil shells were found in blocks of stone carried to the city of Verona to repair the Citadel of San Felice. A number of learned men were consulted about the fossils, including Hieronymous Fracastoro (1483–1553), who, after being Professor of Philosophy at Padua to 1503, had returned to his native city of Verona to practise medicine, and became physician to Pope Paul III. After seeing the shells, Fracastoro decided that they were the remains of animals which had once lived in the place they had been unearthed. He argued with vigour against those who claimed that the shells were emplaced during the Noachian Flood. He thought the Flood too transient an event, consisting largely of fluviatile waters, which would strew the shells over the land surface instead of burying them inside mountains where they had been uncovered by quarrying. Lyell panegyrizes Fracastoro, and clearly feels that geology would have progressed far more rapidly if his views had been heeded:

the clear and philosophical views of Fracastoro were disregarded, and the talent and argumentative powers of the learned were doomed for three centuries to be wasted in the discussion of those two simple and preliminary questions: first, whether fossil remains had ever belonged to living creatures [Fracastoro thought that they did]; and, secondly, whether, if this be admitted, all the phenomena could be explained by the Noachian deluge. (Lyell 1834: i. 34)

A diluvial origin of Italian sea shells

Fracastoro's pronouncement notwithstanding, the lively debate over the nature of fossil organic remains continued, with many scholars supporting the diluvial view. Little of any substance was added to the debate, most writers being content to state that the

Flood had occurred. This is evident from Lyell's summary of the chief Italian works on fossils written in the second half of the sixteenth century (1834: i. 37–8). Girolamo Cardano (1501–76), a physician, mathematician, philosopher, and astrologer, in his *De subtilitate* of 1550, simply expressed the view that fossil marine shells are evidence of the former residence of the sea upon the mountains. Similarly, the celebrated botanist, Andrea Cesalpino, contended, in his *De metallicis* of 1596, that fossil shells had been left on the land as the sea retired. Fresh ideas were occasionally thrown in by members of the anti-diluvial faction. Simeone Majoli, in his *Dies caniculares* of 1597, proposed that the shells found at Verona and at other places might have been cast upon the land by volcanic explosions (Geikie 1905: 53). In France, Bernard Palissy (1510?–89), The Aquitainian potter and enameller who died in the Bastille, having been imprisoned for his religious beliefs, took issue with the Italian diluvialists. His investigations of the geology of the Paris Basin led him to conclude, in his *Discours admirables de la nature des eau et founteines tant naturelles qu'artificelles, des metaus, des sels et salines* of 1580, that not all fossil shells had been strewn by a universal deluge.

The debate on the origin of fossil sea shells found in the Italian mountains continued into the seventeenth century. Some advances were made during this period. Fabio Colonna (1567–1645?), a Neapolitan scholar, botanist, and staunch diluvialist, recognized, in his *Osservazioni sugli animali aquatici et terrestri* of 1616, that some of the fossil sea shells were of marine forms, whereas others were terrestrial forms (Lyell 1834: i. 38–9). But it was more common for the old arguments to be dug up again, claims and counter-claims echoing the views expressed a century before. In a work on the fossils of Calabria, published in Naples in 1670 and entitled *La vana speculazione disingannata dal senso, lettera risponsiva circa i corpi marini che petrificata si trovano in varii luoghi terrestri*, the Sicilian painter, Agostino Scilla, maintained that all fossil shells are the effects and proofs of the Noachian Flood. Joannis Quirini, in his *De testaceis fossilibus* of 1676, countered this view by contending that the Flood waters could not have carried heavy bodies to the tops of mountains, since the agitation of the sea never extended to great depths, a fact which had been demonstrated a few years before by Robert Boyle (see Boyle 1772: iii. 352–4). Lyell (1834: i. 44) records that Quirini was

the first writer who ventured to maintain that the universality of
the Noachian Cataclysm ought not to be insisted upon.

Steno's contribution to the sea-shell debate

It was at this juncture in the debate that an attempt was made by
the Dane, Nicolaus Steno (alias Niels Steensen, 1638–86), to
explain in detail the development of the Tuscan landscape.
Although Steno thought that the Flood had played a prominent
role in the geological history of Tuscany, his thesis was firmly
based on field observation, and is a far more detailed and
penetrating study than the rather routine and uninspiring studies
that most diluvialists were making at the time. While carrying out
his duties as court physician to Grand Duke Ferdinand II at
Florence, Steno, who had always evinced a keen interest in
science, explored the Tuscan landscape. His field observations led
to the publication of a famous treatise, his *Prodomus* of 1669, in
which he explains how fossils came to be entombed in rock, lays
down in sketchy form the principles of stratigraphy, and discusses
the origin of mountains. At the conclusion of his work, Steno
describes the sequence of events which have produced the present
plains and hills in Tuscany. In doing so, he provides a system of
Earth history in which the Flood plays an important, but not a
solo, role. He recognizes six stages, or aspects as he styles them, in
the development of the Tuscan region. Firstly, just after Creation,
the region was covered by a 'watery fluid' out of which inorganic
sediments precipitated to form horizontal, homogeneous strata.
Secondly, the newly formed strata emerged from their watery
covering to form a single, continuous plain of dry land, beneath
which huge caverns were eaten out by the force of fires or water.
Thirdly, some of the caverns might have collapsed to produce
valleys, into which rushed the waters of the Flood. Fourthly, new,
fossiliferous strata of heterogeneous materials were deposited in
the sea which now stood at a higher level than it had prior to the
Flood and occupied the valleys. Fifthly, the new strata emerged
when the Flood waters receded to form a huge plain, and were
then undermined by a second generation of caverns. Finally, the
new strata collapsed into the cavities eaten out beneath them to
produce the Earth's present topography. Clearly, this scheme of
historical geology is a great advance on the simplistic views

of Steno's diluvialist predecessors who saw the Flood as the sole potent force in landscape change.

The sea-shell debate continues

The debate on the Italian fossil shells went on into the eighteenth century. Antonio Vallisnieri (1661–1730), and all later Tuscan geologists, refuted and criticized the cosmogonical systems of Burnet, Whiston, and Woodward (which will be discussed in the next chapter). Vallisnieri carefully observed the marine deposits of Italy and surrounding countries, reporting his findings in a treatise published in Venice in 1721 and entitled *Dei corpi marini, che sui monti si trovano*. He concluded that the sea had once covered the Italian peninsula, a large part of Europe, and, he inferred, the entire globe; and after staying there for a time, had slowly subsided. He maintained that the effects of the sea during its sojourn on land were quite distinct from the effects of the Noachian Flood (Geikie 1905: 60–1). Lyell (1834: i. 59) claims that this opinion, though untenable, was a great advance on Woodward's diluvian hypothesis.

Later, Antonio Lazzaro Moro (1687–1740) suggested in his *Dei crostacei e degli altri marini corpi che si trovano sui monti*, published in Venice in 1740, that the Italian marine deposits could be explained by the action of earthquakes (Geikie 1905: 61–2). He was prompted into making this suggestion after having learnt of a new volcanic island which had appeared in 1707 near Santorini in the Mediterranean. In less than a month, this island had a circumference of half a mile and stood 25 feet above the level of the sea. Before it became covered with volcanic ejecta, the island was found to bear living oysters and crustacea on its surface. Lyell takes delight in relating Moro's ingenious story, designed to ridicule theories then in vogue, the diluvial hypotheses included, of a party of naturalists visiting the island in ignorance of its recent origin:

One [naturalist] immediately points to the marine shells, as proofs of the universal deluge; another argues that they demonstrate the former residence of the sea upon the mountains; a third dismisses them as mere *sports of nature*; while a fourth affirms, that they were born and nourished within the rock in ancient caverns, into which salt water had been raised in the shape of vapour by the action of subterranean heat. (Lyell 1834: i. 60)

However, Moro was not averse to invoking his own brand of sudden and violent events to explain marine shells in mountains, adapting the Mosaic account of Creation to his own system of Earth history. He suggested that on the third day of Creation the Earth consisted of primary rocks, the smooth and regular surface of which was completely covered in water. When God ordained that the dry land should appear, volcanoes erupted, breaking the primary rocks and forming mountains. Melted metals and salts escaped through fissures into the ocean, and the seas gradually became salty. The sand and ashes ejected by the volcanoes were regularly deposited on the ocean bottoms, forming the secondary strata, which in their turn were uplifted by earthquakes. Lyell seems to have been not unfavourably disposed to this part of Moro's systems, at least in that 'upon the whole, it may be remarked, that few of the old cosmological theories had been conceived with so little violation of known analogies' (Lyell 1834: i. 61).

Evidence of the Flood from other countries

It would be wrong to surmise from the above discussion that, where evidence of the Flood was concerned, the Italians had a monopoly. In Switzerland, the diluvial origin of petrified sea animals was championed by the naturalist and ardent fossil and mineral collector, Johann Jacob Scheuchzer (1672–1733). Having been converted to believing that fossil shells were of marine origin by reading Woodward's *Essay*, Scheuchzer set about proving, in his *Piscium querelae et vindicae* of 1708, that the Earth had been remodelled by the Noachian Flood. He had originally regarded fossils as sports of nature, but, having been won over by Woodward's thesis, he affirmed the diluvial distribution of fossils, both in his native Switzerland, and in Europe in general. Geikie notes the quaint humour which runs through Scheuchzer's treatise:

The fossil fishes are represented as assembled in council to protest against their treatment by the descendants of the wicked men that brought on the Flood by which these very fishes had been entombed. They discourse of 'the irrefragible witness of the universal Deluge which by the care of Providence their dumb race places before unbelievers for the conviction of the most daring atheists.' Specimens of their fossil brethren are appealed to—pike, trout, eel, perch, shark—and their well-preserved

minute structure of teeth, bones, scales and fins is pointed to as a
triumphant demonstration that such perfect anatomical detail could be
fabricated by no inorganic process within the rocks, as had been
maliciously affirmed. (Geikie 1905: 100)

Many other eighteenth-century European writers expressed the
belief that the Earth's topography had been remodelled by the
Noachian Flood. However, there would be little advantage in
considering them as their ideas are all very minor variations on the
theme of diluvial metamorphosis. Two of them, Patrick Cockburn
and Alexander Catcott, are more interesting from a geomorphological perspective and will be discussed in the Chapter 5.

4

The Development of Diluvialism
Floods and Restoration cosmogonists

The background to Restoration cosmogony

In the late seventeenth century, several English writers advanced
cosmogonical systems which attempted to explain how the Earth
was formed and how it subsequently developed, in accordance
with the teachings of the Church. These cosmogonies invoked
global floods and cataclysms to explain the history of the Earth.
Later, the promulgators of these cosmogonies were to be derided
and dismissed by the self-proclaimed fathers of modern geology,
from Werner to Playfair and Lyell. Playfair mocks them as unruly
theorists suffering from 'a species of mental derangement, in
which the patient raved continually of comets, deluges, volcanos
and earthquakes; or talked of reclaiming the great wastes of the
chaos, and converting them into a terraqueous and habitable
globe' (Playfair 1811: 207–8). Lyell quotes Voltaire's 'Dissertation
envoyée a l'Académie de Boulogne, sur les changemens arrivés
dans notre globe':

In allusion to the theories of Burnet, Woodward, and other physico-
theological writers, he [Voltaire] declared that they were as fond of
changes of scene on the face of the globe, as were the populace at a play.
'Every one of them destroys and renovates the earth after his own fashion,
as Descartes framed it: for philosophers put themselves without ceremony
in the place of God, and think to create a universe with a word'. (Lyell
1834: i. 95)

Geikie's invective directed at the late seventeenth-century English
cosmogonists makes a similar point. It rebukes them for pro-
mulgating monstrous doctrines about Earth history at a time when
science had not progressed far enough to provide a firm basis for
such speculations, and for supplying the missing data on the
structure of the Earth with wholly imaginary pictures of the history
of Creation; and, it accuses them of obstructing the progress of

geological enquiry by diverting attention from the observation of Nature into barren controversy about speculations (Geikie 1905: 66).

The chief reason why the early nineteenth-century geologists turned so viciously on their scientific progenitors, is that they believed that to find the path towards the true science of the Earth, they needed to start out afresh, on the new track of sedulous observation and disciplined induction: the earlier, armchair theories were ramshackle affairs with no substance, and an unbefitting basis for the rising edifice of hard geological facts (cf. Porter 1977: 1). Such a view is, to some extent, understandable. What is more difficult to fathom is why virtually all subsequent historians of geology have looked so approvingly upon the judgement passed on the late seventeenth-century cosmogonists by Playfair and other geologists at the start of the nineteenth century. After all, the late seventeenth-century cosmogonists were by no means crackpots or cranks: they were some of the intellectual giants of their day; their theories were much admired, had a significant impact on the development of scientific thought, and were the product of a sparkling age of British science. Recently, Porter has come to the defence of these much-slated cosmogonists. He argues that if their theories proved unable to solve the problems of Earth history, it was not through stupidity, but a result of the immanent difficulties and contradictory demands made by their philosophies of nature. The most common tilt made at these unfortunate cosmogonists is that their theories are not the results of solid empirical endeavour, but 'amalgams of imagination and speculation, *a priori* metaphysical and natural philosophical notions, religious Revelation and uncritically re-gurgitated ancient wisdom, derived from books, sanctified by age and authority' (Porter 1977: 63). The counter case to these admonitions, so carefully argued by Porter, is too complicated to present fully here. Nor, in the present context, is it necessary to do so. Suffice it to say that they do contain an element of empiricism, that they were constructed after the fashion of the time which accepted a range of modes and sources of enquiry after truth, and that they were not attempts to tailor nature to fit the Scriptures. They were important, much admired, much satirized, much criticized products of their own age, and it seems fairer to view them in that context, than as stumbling-blocks on the path of geological progress.

The Flood and a vengeful God

Thomas Burnet (1635–1715), an English author and clergyman, published in 1681 under the patronage of Charles II a book, eloquently written in Latin, and entitled *Telluris theoria sacra*. With the encouragement of King Charles II, the book was translated into English and published in 1684 as *The sacred theory of the Earth: Containing an account of the original of the Earth, and of all the general changes which it hath already undergone, or is to undergo till the consummation of all things*. The English version ran into at least six editions, the second edition of 1691 being reprinted in 1965.

Burnet's book is full of vivid imagery and eloquent prose. The theory it contains is Burnet's attempt to explain how there could have been enough water in the Flood to cover the highest mountains, a problem which had vexed scholars for a long time. Burnet's solution is to propose that before the Flood, the Earth's surface had a totally different configuration from the present one; that it was, in fact, smooth, regular, and uniform, like the surface of an egg. Burnet explains that the flooding of this featureless world would require no more water than is contained in the present oceans.

Burnet conjectures how the featureless, antediluvian world came into being. He proposes that the Earth was originally, six thousands years ago, a chaotic liquid mixture of earth, water, oil, and air which gradually consolidated to form a sphere. As time passed, the rocky ingredients separated out from the chaotic fluid. The heaviest material in the liquid fell and collected at the Earth's centre where it formed a spherical core. The next heaviest portions of the chaotic fluid then became the terrestrial fluids, while the least heavy portions became the atmosphere. The terrestrial fluids further separated, oily, fatty, and light fluids rising to the surface to float on underlying water. Further separation also took place in the atmosphere, which was then thick and dark owing to the suspension of terrestrial particles. Slowly, the terrestrial particles settled out and mixed with the fatty and oily materials floating on the water to form a hard, congealed skin lying on the surface of the terrestrial fluids and completely sealing them in a watery abyss. The congealed superficial layer provided sustenance for the first animals, plants, and humans who lived in this antediluvian paradise. There were at this time no mountains, no seas, no storms, no rainbows; the Earth's axis of rotation was normal to the orbital plane so there was perpetual equinox with no seasons.

Burnet maintains that the paradisiacal, antediluvian period lasted 1600 years. Then, owing to man's wickedness, God engineered a cataclysm. The forty days and nights of rain mentioned in Genesis were, according to Burnet, insufficient to cause the Flood. He believed that the true cause of the Flood was the widening and deepening of cracks which already existed in the Earth's smooth surface by God's direct intervention. This divine action caused the release of waters from the watery abyss, which then surged over the globe as tidal waves so massive that, if it had not been for God's help, the ark would have sunk. Here is how Burnet describes the Flood itself:

let us suppose the great frame of the exteriour Earth to have broke at this time, or the Fountains of the great Abysse, as *Moses* saith, to have been then open'd, from thence would issue, upon the fall of the Earth, with an unspeakable violence, such a Flood of waters as would over-run and overwhelm for a time all those fragments which the Earth broke into, and bury in one common Grave all Mankind, and all the Inhabitants of the Earth. Besides, if the *Flood-gates* of Heaven were any thing distinct from the Forty days Rain, their effusion, 'tis likely, was at this same time when the Abysse was broken open; for the sinking of the Earth would make an extraordinary convulsion of the Regions of the Air, and that crack and noise that must be in the falling World, and in the collision of the Earth and the Abysse, would make a great and universal Concussion above, which things together, must needs so shake, or so squeeze the Atmosphere, as to bring down all the remaining Vapours; But the force of these motions not being equal throughout the whole Air, but drawing or pressing more in some places than in other, where the Centre of the convulsion was, there would be the chiefest collection, and there would fall, not showers of Rain, or single drops, but great spouts or caskades of water; and this is that which *Moses* seems to call, not improperly, the *Cataracts* of Heaven, or the *Windows of Heaven being set open*. (Burnet 1965: 83–4)

Burnet considers the shock of the sudden release of waters from the watery abyss to have been so great that the earth shook on its axis, which shifted to its present tilt. Burnet then describes how some fragments of the shell sank into the abyss, crumbling as they did so to form mountains, valleys, and islands; and in some places the foundering of the Earth's shell was so extreme that the crust had disappeared from view, leaving the abyssal waters exposed. To explain why the Flood waters receded, Burnet argues that within the Earth were great caverns into which the Flood waters could drain, so allowing the reappearance of dry land. And so it was, in Burnet's interpretation, that the Earth became a wasteland, a dirty little planet, a befitting abode for man the sinner.

Although the catastrophe associated with the fall of man was identified by Burnet as the biblical Flood, and although Burnet professed that his theories matched the events described in Genesis, he was, as Chorley *et al.* (1964: 11) report, ridiculed in a popular ballad as saying

> That all the books of Moses
> Were nothing but supposes.

Burnet's thesis was contested by the mathematician John Keill (1698), who questioned, among many other things, the suggestion that the Earth had shifted on its axis. But, its critics notwithstanding, *The sacred theory of the Earth* was an immensely popular book, even if, as was widely realized in Burnet's day, it was little more than a theologian's reworking of the system proposed by the great French philosopher, René Descartes, in his *Principia philosophiae* of 1644. Indeed, most of the ingredients of all cosmogonical systems, starting with Burnet's, are present, sometimes in a different guise, in Descartes' system: the origin of the Earth as a hot, fiery ball; the collapse of the Earth's outer crust; the release of massive volumes of water. It is fair to say that Descartes, as well as having had a profound influence on scientific thought, also had a fundamental influence on theories of Earth history.

Flood waters from the Abyss

Three possible causes of the Flood

A number of Restoration writers, inspired by Burnet, produced elaborate and logical cosmogonies which explained the Flood by assuming that its waters had originally been housed in enormous cavities within the Earth. These subterranean waters were then, for various reasons, released and issued on to the surface of the globe, causing deep and widespread inundation. The first writer to follow Burnet's lead was the famous botanist and zoologist, John Ray (1628–1705). Ray expounded his cosmogonical system in his two very popular works, *The wisdom of God manifested in the works of Creation* (1691), and *Miscellaneous discourses concerning the dissolution and changes of the world* (1692), called *Three physico-theological discourses* (1693) in later editions. (The *Three physico-theological discourses* deal with the Creation, the Deluge, and the final Conflagration respectively.)

Ray discussed the Flood in considerable detail. He appears to have had an open mind as to the source of the Flood waters, and suggested three possibilities. One possibility was that the Earth's centre of gravity had gradually shifted, bringing it nearer to the eastern hemisphere and then returning it to its original location. This process would lead to the inundation of Asia, but would leave America dry. Another possibility was that a pressure upon the surface of the Atlantic and Pacific oceans forced the waters down into the abyss and out upon the land through cracks in the Earth's crust. By this process, the discovery of marine organisms on the land could be explained, the shells being carried with the ocean water through the abyss and out at the fissures. A remote possibility was that the Flood waters could have come from the air. Ray made elaborate calculations, using data on annual rainfall, of the floods caused by occasional thunderstorms, and the discharge of river water into the sea, which showed that rain could produce twenty times eighty oceans of water.

A liquefying Flood

Another writer to elaborate on the subterranean origin of the Flood was John Woodward (1665–1728), a professor of physick at Gresham College and a member of the Royal Society. Woodward travelled 'underground' England in search of fossils and rock formations, visiting caves, collieries, and quarries. He was, evidently, the first person fully to appreciate the vital importance of fieldwork in geological studies (Davies 1969: 75). In his *An essay towards a natural history of the Earth*, published in 1695, Woodward observed that many fossil assemblages in England contain representatives from diverse parts of the globe and commonly involve a curious mixture of marine and terrestrial species. He concluded that a mere change in the distribution of land and sea could not explain the juxtaposition of these fossils. To explain the puzzling mixture of native and exotic species, he invoked a cataclysmic event—the biblical Flood. He proposed that 'these marine bodies were born forth of the sea by the universal deluge: and that, upon the return of the water back again from off the Earth, they were left behind at land' (Woodward 1695: 72). He accounted for the Flood by arguing that

there is a mighty *collection of water* inclosed in the *bowels of the Earth*, constituting an *huge orb* in the interiour or central parts of it; upon the surface of which orb of water the terrestrial *strata* are expanded . . . this is the same which *Moses* calls the *Great Deep*, or *Abyss*: the ancient gentile writers, *Erebus*, and *Tartarus*. (Woodward 1695: 117)

The waters from the abyss were, by God's will, released and the whole globe taken to pieces and dissolved. All the rocks, minerals, and animal and vegetable bodies were liquefied in the Flood waters, save the antediluvian animals and plants which for some reason were immune to dissolution:

during the time of the Deluge, whilst the water was out upon, and covered the terrestrial globe, all the stone and marble of the antediluvian Earth: all the metalls of it: all mineral concretions: and, in a word, all fossils whatever that had before obtained any solidity, were totally dissolved, and their constituent corpuscles all disjoyned, their cohaesion perfectly ceasing. (Woodward 1695: 74)

The turbid liquid held the contents of the antediluvian world in solution and suspension:

the said corpuscles of these solid fossils, together with the corpuscles of those which were not before solid, such as sand, earth, and the like: as also all animal bodies, and parts of animals, bones, teeth, shells: vegetables, and parts of vegetables, trees, shrubs, herbs: and, to be short, all bodies whatsoever that were either upon the Earth, or that constituted the mass of it, if not quite down to the Abyss, yet at least to the greatest depth we ever dig: I say all these were assumed up promiscuously into the water, and sustained in it, in such manner that the water, and bodies in it, together made up one common confused mass. (Woodward 1695: 74–5)

At length, the contents of the liquid were released and sank down to the Earth's centre. The materials sank in an ordered sequence, according to their specific gravity, to form horizontal layers or strata. The heaviest elements fell first forming the lowermost strata, and the lighter elements fell afterwards forming the uppermost strata:

that matter, body, or bodies, which had the greatest quantity or degree of gravity, subsiding first in order, and falling lowest: *that* which had the next, or a still lesser degree of gravity, subsiding next after, and settling upon the precedent: and so on in their several courses: *that* which had the least gravity sinking not down till last of all, settling at the surface of the sediment, and covering all the rest. (Woodward 1695: 75)

The plant and animal remains, which had proved resistant to the solvent action of the Flood waters, also settled according to their specific gravity, the heavier ones thus being lodged in the heavier parent strata, the light ones in the lighter parent strata:

the shells of those cockles, escalops, perewinkles, and the rest, which have a greater degree of gravity, were enclosed and lodged in the *strata* of stone, marble, and the heavier kinds of terrestrial matter: the lighter shells not sinking down till afterwards, and so falling amongst the lighter matter, such as chalk, and the like, in all such parts of the mass where there happened to be any considerable quantity of chalk, or other matter lighter than stone; but where there was *none*, the said shells fell upon, or near unto, the surface; . . . humane bodies, the bodies of quadrupeds, and other land animals, of birds, of fishes, both of the cartilaginous, squamose, and crustaceous kinds; the bones, teeth, horns, and other parts of beasts, and of fishes: the shells of land snails: and the shells of those river and sea shell-fish that were lighter than chalk &c. Trees, shrubs, and all other vegetables, and the seeds of them: and that peculiar terrestrial matter whereof these consist, and out of which they are all formed: I say all these . . . being, bulk for bulk, lighter than sand, marl, chalk, or the other ordinary matter of the globe, were not precipitated till the last, and so lay above all the former, constituting the supreme or outmost *stratum* of the globe. (Woodward 1695: 76–8)

Woodward was explicit that all the strata were deposited as large horizontal sheets:

the said *strata*, whether of stone, of chalk, of cole, of earth, or whatever other matter they consisted of, lying thus upon each other, were all originally parallel . . . they were plain, eaven, and regular; and the surface of the Earth likewise [was] even and spherical: . . . they were continuous, and not interrupted, or broken: . . . the whole mass of the water lay then above them all, and constituted a fluid sphere environing the globe. (Woodward 1695: 79–80)

Later, the strata were subjected to Earth movements, the cause of which Woodward merely describes as being seated within the Earth: 'after some time the *strata* were broken, on all sides of the globe: . . . they were dislocated, and their situation varied, being elevated in some places, and depressed in others' (Woodward 1695: 80). The effect of the disruption and dislocation of the strata was to produce the present topography of the Earth:

all the irregularities and inequalities of the terrestrial globe were caused by this means: date their original from this disruption, and are all entirely owing unto it. . . . In one word, . . . the *whole terraqueous globe* was, by this means, at the time of the Deluge, put into the condition that we now behold. (Woodward 1695; 80–1)

Lyell concludes of Woodward that, 'in his anxiety to accommodate all observed phenomena to the scriptural account of the Creation and Deluge, he arrived at most erroneous results' (Lyell 1834; i. 53). Chorley *et al.* (1964) are more charitable, explaining that although 'the argument contains obvious errors, for the period in which it was made it represents a genuinely intelligent explanation and had the prime merit of avoiding conflict with the Church' (Chorley *et al.* 1964; 13). Woodward himself no doubt sincerely believed that his interpretation of Earth history was entirely dispassionate, and based only upon the evidence he saw in the field; but in fact 'his mind was so conditioned by the bibliolatry of the age that he achieved nothing more than the examination of Nature through a pair of Mosaic spectacles' (Davies 1969; 77).

One of the most striking aspects of Woodward's work is his conception of God, so utterly different from Burnet's conception of a decade before:

Burnet adhered to the older, Puritan view of God, and regarded Him as mankind's angry, vengeful judge who had ordained that there should be progressive, punitive decay of the entire universe. He saw the Deluge as merely one terrible episode in this universal degeneration of Nature, and he believed that man was condemned to eke out his miserable present existence amidst the grotesque ruins of the magnificent ante-diluvial world which his ancestors had so grossly misused. Woodward, on the other hand . . . lived at the dawn of a new era. After the Restoration, the old Puritan conception of the Deity rapidly faded, and with it there passed the belief in a decadent universe. In place of the wrathful God of the Puritans, there emerged a new conception of the Deity as a benign, compassionate Being who was the architect of a magnificent Creation. This was the view to which Woodward subscribed. Like many of his contemporaries he leaned strongly towards a teleological interpretation of Nature, and because of this he felt impelled to seek the divine benevolence even in the tumult of the Flood. He felt it inconceivable that a merciful God could have used the Flood solely as a vicious, destructive vehicle for the divine vengeance, and he firmly rejected both Burnet's belief that the Flood had been an unmitigated catastrophe, and his claim that the present Earth is a chaotic ruin utterly devoid of plan. (Davies 1969: 79–80)

Flood waters from the tail of a comet

William Whiston (1666–1753), an English mathematician and protégé of Isaac Newton, developed his cosmogony in his popular book with the long title *A new theory of the Earth, from its original, to the consummation of all things: Wherein the creation of the world in six days, the universal deluge, and the general conflagration, as laid down in the Holy Scriptures, are shewn to be perfectly agreeable to reason and philosophy. With a large introductory discourse concerning the genuine nature, stile, and extent of the* Mosaick *history of the Creation* (1696). The book itself is long, and tightly argued. In the large introductory discourse, Whiston sets down his basic proposition:

The Mosaick *Creation is not a nice and philosophical account of the origin of all things; but an historical and true representation of the formation of our single Earth out of a confused chaos, and of the successive and visible changes thereof each day, till it became the habitation of Mankind.* (Whiston 1696: 3)

Later in the introduction, he establishes three postulata:

I. The obvious or literal sense of Scripture is the true and real one, where no evident reason can be given to the contrary.

II. That which is clearly accountable in a natural way, is not without reason to be ascribed to a Miraculous power.

III. What ancient tradition asserts of the constitution of Nature, or of the origin and primitive states of the world, is to be allow'd for true, where 'tis fully agreeable to Scripture, reason, and philosophy. (Whiston 1696: 95)

The rest of the treatise, which comprises five books, amplifies these basic propositions. The resulting cosmogony, which leans heavily upon on the events described in Genesis, is like Burnet's cosmogony in many respects save one: Whiston believed that Burnet's cosmogony overlooked the role of comets in the history of the Earth. As Lyell explains,

The remarkable comet of 1680 was fresh in the memory of every one when Whiston first began his cosmological studies, and the principal novelty of his speculations consisted in attributing the deluge to the near approach to the earth of one of these erratic bodies. (Lyell 1834; i. 56)

Whiston proposed that comets are a species of planet which revolve about the Sun (lemma xlii, p. 36), in very oblong and eccentric elliptical orbits (lemma xliii, p. 36), and may pass through the planetary system (lemma xlvi, p. 37). It followed from these propositions that '*we may observe a new possible cause of fast changes in the planetary world, by the access and approach of these vast and hitherto little known bodies to any of the planets*' (Whiston 1696: 37). Whiston hypothesized that the Earth had originally been a comet: 'The ancient *Chaos*, the origin of our Earth, was the atmosphere of a comet' (Whiston 1696: Hypothesis I, p. 69). On nearing the Sun, the proto-Earth was melted to form a coherent mass. On moving away from the Sun, the terrestrial materials became rearranged, the heavier materials forming a solid core, the lighter materials collecting to form the superficial parts. In its antediluvian state, the Earth was covered by water, save for high mountain chains and islands which stood above the oceans. On the antediluvian Earth,

The ancient Paradise or Garden of *Eden*, the seat of our first parents in the State of Innocence, was at the joynt course of the rivers *Tigris* and *Euphrates*; either before they fall into the *Persian Gulf*, where they now unite together, and separate again; or rather where they anciently divided themselves below the island *Ormus*, where the *Persian Gulf*, under the Tropick of *Cancer*, falls into the *Persian Sea*. (Whiston 1696: Hypothesis IV, p. 104)

And 'The primitive ecliptick, or its correspondent circle on the Earth, intersected the present Tropick of *Cancer* at Paradise; or at least at its meridian' (Whiston 1696: Hypothesis V, p. 106). Then, owing to man's sinning, God caused a universal Deluge using a comet as his instrument: 'A comet, descending, in the plain of the ecliptick, towards its *perihelion*; on the first day of the Deluge past just before the body of our Earth' (Whiston 1696: Hypothesis X, p. 126).

Whiston, in the *New theory of the Earth*, in an appendix to the third edition of that work, and in his *Astronomical principles of religion, natural and reveal'd* (1717), describes several phenomena relating to the universal Deluge, and its effect upon the Earth. These phenomena are worth considering here. Whiston dates the Deluge as commencing in the seventeenth century from the Creation, on Thursday 27 November in the year 2349 BC. The prodigious amount of water in the Deluge was occasioned by a most extraordinary and violent rain, which fell without stopping for forty days and forty nights. After a brief intermission, the rains fell again and continued to do so for a hundred and fifty days after the Deluge had begun. However, the source of this superabundant rainfall was not the Earth, or the seas, but, in large part, the vapours in the tail of a passing comet, and, in small part, water released from the central Abyss. Whether the Flood waters were calm or stormy was unclear: Whiston found it difficult to reconcile the fact that Noah's ark would have been unable to abide a stormy sea, which suggests that the waters of the Deluge were calm, with the fact that the Scriptures describe violent winds and storms during the Flood. Calm or stormy, the Flood waters increased little by little till they attained their utmost height, fifteen cubits above the highest mountains. They then subsided by degrees, being evaporated by wind and descending through fissures into the bowels of the Earth, till they disappeared from the Earth's face, leaving the present continents. Whiston suggested that most of the passages leading to the abyss were found in mountainous areas, and, since these areas would drain first, the drainage of low-lying areas and the oceans would be slow. The abyss could not hold all the Flood waters, and the remaining portion formed the present oceans. The Flood was universal and destroyed all the land animals save those housed in the ark. To Whiston, unlike Burnet, the Flood was both a signal instance of Divine vengeance on a wicked world, and the effect of the peculiar and extraordinary providence of God (Whiston 1696; 207). The passage of the comet also caused the Earth to start turning about its axis, and to adopt an elliptical orbit round the Sun:

Tho' the *annual motion* of the Earth commenc'd at the beginning of the *Mosaick* Creation; yet its *diurnal rotation* did not till after the Fall of Man. (Whiston 1696: Hypothesis III, p. 79)

The original orbits of the planets, and particularly of the Earth, before the Deluge, were perfect circles. (Whiston 1696: Hypothesis VII, p. 110)

Whiston's contemporaries were divided over the merit of his treatise. Ray deemed it rather odd and extravagant, and John Keill (1698) disliked it, arguing with force that it could be reconciled with neither Moses nor physical science; on the other hand, both John Locke and Isaac Newton applauded it (Force 1983: 8–9). Later commentators have generally been scathing of Whiston's work. Lyell writes,

He [Whiston] had the art to throw an air of plausibility over the most improbable parts of his theory, and seemed to be proceeding in the most sober manner, and by the aid of mathematical demonstration, to the establishment of his various propositions. . . . Like all who introduced purely hypothetical causes to account for natural phenomena, Whiston retarded the progress of truth, diverting men from the investigation of the laws of sublunary nature, and inducing them to waste time in speculations on the power of comets to drag the waters of the ocean over the land—on the condensation of the vapours of their tails into water, and other matters equally edifying. (Lyell 1834; i. 56–7)

Floods and earthquakes

A seminal treatise was penned by the English scientist, Robert Hooke (1635–1703). Hooke's writings, many of which were published posthumously (Waller 1705), show that in matters of Earth history he had an uncommon perspicacity. He never constructed a true cosmogony, but his views on fossils and changes of land and sea are highly advanced. In his 'Discourse on earthquakes', which he completed in 1688, Hooke expresses the opinion that fossils of unknown forms of animals and plants are the remains of extinct species, but that the Flood was of too short a duration to account for all the world's fossiliferous strata. Asking himself how the present areas of land came to be dry, he answers

it could not be from the Flood of *Noah*, since the duration of that which was but about two hundred natural days, or half an year could not afford time enough for the production and perfection of so many and so great and full grown shells, as these which are so found do testify; besides the quantity and thickness of the beds of sand with which they are many times found mixed, do argue that there must needs be a much longer time of the seas residence above the same, than so short a space can afford. (Hooke 1688; 341)

Nor does he think that a gradual swelling of the Earth could explain the distribution of dry land. He contests that the present dry lands could not

proceed from a gradual swelling of the Earth, from a subterraneous fermentation, which by degrees should raise the parts of the sea above the surface thereof; since if it had been that way, these shells would have been found only at the top of the Earth or very near it, and not buried at so great a depth under it as the instances I mentioned of the layer of shells in the *Alps* buried under so vast a mountain, and that near the *Needles* in the *Isle of Wight* found in the middle of an hill, could not rationally be so caused. (Hooke 1688; 341–2)

Instead, he suggests that the presence of such fossils on continents is unequivocal evidence that the distribution of land and sea has changed greatly and catastrophically, owing to the agency of earthquakes. He uses the evidence of fossil sea shells on land to support his proposition

That a great part of the surface of the Earth hath been since the Creation transformed and made of another nature; namely, many parts which have been sea are now land, and divers other parts are now sea which were once a firm land; mountains have been turned into plains, and plains into mountains, and the like. (Hooke 1688: 290)

As to the role earthquakes play in this interchange, he is quite specific, enumerating four chief effects:

The first is the raising of the superficial parts of the Earth above their former level: and under this head there are four species. The 1st is the raising of a considerable part of a country, which before lay level with the sea, and making it lye many feet, nay, sometimes many fathoms above its former height. A 2nd is the raising of a considerable part of the bottom of the sea, and making it lye above the surface of the water, by which means divers islands have been generated and produced. A 3rd species is the raising of very considerable mountains out of a plain and level country. And a 4th species is the raising of the parts of the Earth by the throwing on of a great access of new earth, and for burying the former surface under a covering of new earth many fathoms thick.

A second sort of effects perform'd by earthquakes, is the depression or sinking of the parts of the Earth's surface below the former level. Under this head are also comprised four distinct species, which are directly contrary to the four last named.

The *first*, is a sinking of some parts of the surface of the Earth, lying a good way within the land, and converting it into a lake of almost unmeasurable depth.

The *second*, is the sinking of a considerable part of the plain land, near the sea, below its former level, and so suffering the sea to come in and overflow it, being laid lower than the surface of the next adjoining sea.

A *third*, is the sinking of the parts of the bottom of the sea much lower, and creating therein vast *vorages* and *abysses*.

A *fourth*, is the making bare, or uncovering of divers parts of the Earth, which were before a good way below the surface; and this either by suddenly throwing away these upper parts by some subterraneous motion, or else by washing them away by some kind of eruption of waters from unusual places, vomited out by some earthquake.

A third sort of effects produced by earthquakes, are the subversions, conversions, and transpositions of the parts of the Earth.

A fourth sort of *effects*, are *liquefaction, baking, calcining, petrifaction, transformation, sublimation, distillation,* &c. (Hooke 1688; 298–9)

Thus Hooke envisaged the sudden subsidence of prediluvian land masses to form the present ocean basins, and the sudden upheaval of pristine ocean floors, with their accumulated banks of sea shells, to form the present continents, all owing to the agency of earthquakes.

Without doubt, Hooke's treatise stands as a masterpiece. Lyell (1834: i. 46) regards it as the most philosophical production of its age concerning the causes of former changes in the organic and inorganic worlds. It is certainly one of Restoration England's finest productions, and is in many ways more enlightening than the writings of the many eighteenth-century philosophers who, as will now be seen, proposed yet more systems of Earth history.

5

The Maturing of Diluvialism
Floods and Enlightenment philosophers

It is commonly believed that the eighteenth century saw something of a hiatus in the development of Earth science. Davies (1969: 95) diagnoses this gap as a symptom of a general lethargy which overtook European science soon after 1700. He detects a revival of interest in England in 1778 with the publication of John Whitehurst's *An enquiry into the original state and formation of the Earth*, and abroad with the writings of men such as Desmarest, Guettard, Lehmann, Pallas, de Saussure, and Werner. Eyles (1969) recognizes a tendency among modern geologists to regard the writings of their eighteenth-century colleagues as little more than 'wild theorizing and fruitless speculation'. However, he argues that, on close inspection of the material, this view cannot be maintained. Certainly, during the early eighteenth century, the works of late seventeenth-century writers were, judging by the number of translations, editions, and reprints which appeared, widely read and discussed in Britain and throughout much of Europe. Burnet's *Sacred theory of the Earth* was reprinted at least six times during the eighteenth century, the latest appearing in 1759 (Eyles 1969: 163). Several English editions of Woodward's *Essay towards a natural history of the Earth* were published during the eighteenth century, and it was translated into Latin, French, Italian, and German. At least six English editions of Whiston's *New theory of the Earth* appeared during the eighteenth century, the last in 1755; and a German edition was published in 1713. The works of John Ray, written towards the close of the seventeenth century, were also frequently reprinted and translated in the first half of the eighteenth century. New works also appeared. Midway through the eighteenth century, new systems of Earth history were published. Geikie regards these new cosmogonies as of a very different stamp from the ones vended by Burnet, Whiston, Ray, and Woodward. He believes they were written by 'men who took a broad view of the world and endeavoured to trace its origin and progress in the light of what was then known of the laws of Nature' (Geikie 1905: 79). The new cosmogonies were written by writers from mainland Europe, and all, save that put forward by Leibnitz, were a product of the European Enlightenment which had begun in the late seventeenth century.

Enlightened cosmogonies from France

A slowly retiring, universal sea

One of the first of the new generation of cosmogonies was, at its author's request, published posthumously in 1748. Benoit de Maillet (1656–1738) was a French diplomat and traveller, and a 'keen and shrewd observer of nature' (Geikie 1905: 84). During his life, he acquired considerable first-hand experience of the geology and historical changes in the countries surrounding the Mediterranean Sea, on the basis of which he constructed a cosmogonical system. He deemed his views too unorthodox to make them public during his lifetime, and even in posthumous publication he chose to hide his identity by writing under the guise of an Indian philosopher called Telliamed, which is, of course, de Maillet spelt backwards. Telliamed's book is called *Telliamed ou entretiens, sur la diminution de la mer, d'un philosophe Indien avec un missionaire Français* (1748). As the title suggests, it takes the form of a dialogue between an Indian philosopher and a French missionary. The chief argument of the book is that the Earth was once wholly enveloped in water. Gradually, the water was diminished, and will continue to diminish until the planet is dry, when it will be engulfed in a conflagration fuelled by the outbreak of volcanic activity. De Maillet sees the Earth as a product of the sea: mountains consist of sediments formed by the sea, the oldest and loftiest of which are made of a simple and uniform substance in which few or no traces of animal life have been preserved. When the sea level had diminished enough to expose the tops of the earliest mountains, waves pounded their flanks and in doing so produced sediment from which new mountains could be made. That these sediments are laid in layers is to be expected from the present action of the sea along its coast and on its bottom. Organic remains become increasingly abundant in the newer mountains.

De Maillet lays considerable stress on the marine shells found on mountain tops as evidence of the former covering of water. He finds it impossible to believe that universal marine formations (strata) were deposited by the Noachian Deluge, which he considers to be a local and transient inundation. The valleys and other hollows of the Earth's surface, he claims, were scooped out by marine currents during the subsidence of the sea, leaving the mountain ridges standing between them. The gradual diminution of the ocean waters takes place by evaporation, the water vapour being lost to space.

Epochs of Nature and a gradually cooling planet

Georges Louis Leclerc, Comte de Buffon (1708–88), was one of the great *philosophes* of the Enlightenment. Geikie is full of praise for his contribution to Earth science:

Endowed with a spirit of bold generalisation, and gifted with a style of singular clearness and eloquence, he was peculiarly fitted to fascinate his countrymen, and to exercise a powerful influence on the scientific progress of his age. He is the central figure in a striking group of writers and observers who placed France in the very front of the onward march of science, and who laid some of the foundation-stones of modern geology. . . . In breadth and grandeur of conception Buffon far surpassed the earlier writers who had promulgated theories of the earth. (Geikie 1905: 88–9, 96)

Buffon expounded his grandiose and ingenious theory of the Earth in his *magnum opus* entitled *Histoire naturelle, générale et particulière, avec la description du Cabinet du Roi*. The first three volumes of this elephantine work were published in 1749: the first volume contained 'La théorie de la terre' and 'Le système sur la formation des planètes'; the second volume 'L'Histoire générale des animaux' and 'L'Histoire particulière de l'homme'; the third, a 'Description du Cabinet du Roi' (by Daubenton) and a chapter on 'Les variétés de l'espèce humaine'. The next twelve volumes (1755–67) dealt with the history of the quadrupeds. Subsequently, he published in ten volumes 'L'Histoire naturelle des oiseaux et des minéraux' (1771–86), besides seven volumes of 'Suppléments' (1774–89), the most striking of which is the fifth volume, *Les époques de la nature*, a book dated 1778 but not actually published until 1779 and issued as a separate book in two volumes in 1780 (Eyles 1969).

Buffon firmly believed that natural historians should base their theories on observed, commonplace events, and not on extraordinary events such as the passage of comets and the sudden appearance of new planets. He evinced a general disapprobation of the cosmogonists. In particular, he condemned Burnet's *Theory of the Earth* as a 'well written romance, a book which may be read for pleasure, but which ought not to be consulted with a view to instructing oneself' (quoted in Fellows and Milliken 1972: 68). The very notion of a system of Earth history was repugnant to Buffon:

In Buffon's usage in the 1740's the word *théorie* was not a synonym for *hypothesis* or *system*, both anathema to him at that time; it quite specifically designated 'a close induction from established facts.' He clearly felt himself to be on distinctly different ground from Whiston, Woodward *et al.*, felt indeed that he had succeeded in extracting, and properly ordering, the few kernels of truth embedded in their wild lucubrations. (Fellows and Milliken 1972: 69)

To uncover the history of the Earth, Buffon assumed three observations to be of key importance:

firstly, one finds fossil shells throughout the world; secondly, the materials which make up the Earth are always arranged in horizontal and parallel layers; thirdly, the mountains everywhere show corresponding angles. (Gascar 1983: 101; author's translation)

He concluded from these facts that

the sea formerly covered the Earth completely, laying down on it shells, sediments in successive layers, and by its movement, raising and sculpting the mountains. Originally the result of the original fire, the Earth is, according to Buffon, 'the work of the waters'. (Gascar 1983: 101–2; author's translation)

Buffon commenced his theory of the Earth with the origin of the Solar System. He proposed that the bodies in the Solar System were formed by a collision between the Sun and a comet (an event which must surely be classed as sudden and violent!). The collision led to fiery fragments of molten material from the Sun's surface being hurled into space as 'torrents of matter' which stayed in a heliocentric orbit. One of these fragments became the Earth. Buffon deemed that the later epochs of Earth history followed the

account given in Genesis, though he did not suggest that the six
'days' of the Creation should be taken literally. The first epoch was
one of extreme incandescence during which the Earth remained as
a fiery ball for 2936 years, according to Buffon's calculations. The
second epoch saw the cooling of the Earth with a solidification of
the molten mass, and its crinkling to form primitive mountain
chains. Fellows and Milliken (1972: 69) dub this epoch the 'too-
hot-to-handle stage', because to determine the cooling time of the
terrestrial globe, Buffon used four or five 'pretty young women,
with very soft skin' to hold in turn all sorts of materials which had
been heated red hot, and to tell him the degrees of heating and
cooling! By the third epoch, which started after 30 000 to 35 000
years had elapsed, the globe had cooled enough to permit the
condensation of water vapour from the atmosphere to form a
universal ocean, which stood nine to twelve thousand feet higher
than the present sea level. Buffon believed that such a deep ocean
partly accounted for the presence of marine fossils high in
mountains. It was during this epoch that marine animals and plants
first appeared, but as the seas were then much hotter than they are
today, only heat-tolerant species existed which are now extinct.
During the fourth epoch, the inner parts of the Earth continued to
cool. In places, contraction took place causing cavities to open in
the Earth's surface. Seawater was drained into the subterranean
cavities for about 20 000 years, until the ocean reached its present
level. Volcanoes also began to erupt during this epoch, the
continents appeared, and the present system of valleys was gouged
out by ocean currents. The fifth epoch saw the then warm northern
lands as the home of elephants and other tropical animals. The
sixth epoch saw the continents become divided between the Old
World and New World. Buffon was led to believe that this event
had taken place because the similarity of certain fossils found in
America and Eurasia indicated that the land masses had formerly
been continuous. The seventh and final stage saw the cooling of
the surface and the gradual erosion of higher areas, and, most
importantly to Buffon, the appearance of man, who was created
when the Earth was cool enough for humans to survive.

New cosmogonies from Germany

Floods and a contracting Earth

The first of the new cosmogonies was expounded by the great German mathematician and philosopher Baron Gottfried Wilhelm von Leibnitz (1646–1716). Leibnitz had an interest in Earth history which might have been fostered by his meeting Steno at the court of the Duke of Hanover; certainly, he applauded Steno's *Prodomus* (Eyles 1969: 165). Leibnitz's cosmogony, presented in his *Protogaea*, had been originally published in a much abbreviated, and little read, form in 1693 (Eyles 1969; 165). However, it was not printed in full, and not widely read, until 1749. It is thus best considered in the context of the eighteenth century, even though it was a product of the previous century.

Leibnitz accepted the Cartesian view that matter in the primitive Earth was in a fluid-like state owing to the intense heat, and that the Earth adopted a spherical shape owing to the aggregation of swirling basic building blocks (monads). Whereas Descartes invoked a principle of momentum to explain the further development of the Earth, Leibnitz elicited a dynamical force to separate light from darkness, active elements of the universe from passive elements, and later, to segregate the various inactive elements to form the land and the oceans. Leibnitz's account of Earth history is admirably summarized by Lyell:

He imagined this planet to have been originally a burning luminous mass, which ever since its creation has been undergoing refridgeration. When the outer crust had cooled down sufficiently to allow the vapours to be condensed, they fell, and formed a universal ocean, covering the loftiest mountains, and investing the whole globe. The crust, as it consolidated from a state of fusion, assumed a vesicular and cavernous structure; and being rent in some places, allowed the water to rush into the subterraneous hollows, whereby the level of the primeval ocean was lowered. The breaking in of these vast caverns is supposed to have given rise to the dislocated and deranged position of the strata 'which Steno had described,' and the same disruptions communicated violent movements to the incumbent waters, whence great inundations ensued. The waters, after they had been thus agitated, deposited their sedimentary matter during intervals of quiescence, and hence the various stony and earthy strata. 'We may recognize, therefore,' says Leibnitz, 'a double origin of primitive masses, the one by refridgeration from igneous fusion, the other by concretion from aqueous solution.' By repetition of similar causes (the disruption of the crust and consequent floods), alternations of new strata were produced, until at length these causes were reduced to a condition of quiescent equilibrium, and a more permanent state of things was established. (Lyell 1834: i, 45–6)

Lyell does not make it clear that Leibnitz thought that the floods were produced by the release of water from within the Earth, from the enormous cavities which formed in the crust as it cooled and consolidated. Leibnitz suggested that these cavities broke, owing either to the weight of overlying material or to the explosion of gases, and water issued from them at the Earth's surface to join the water in rivers. The result was vast inundations which deposited sediments over a large area. These sediments then hardened and, when another inundation occurred, a new layer of sediment was deposited on top of them (see Mather and Mason 1939: 45–6).

Von Zittel (1901: 28) thinks that the cosmogonical theory of Leibnitz suffered in the original from a lack of clarity in the wording, and was strained because of the author's desire to construct a history of the Earth which was in accord with the Mosaic Creation. None the less, it stands on a par with the cosmogonies of the English quartet—Burnet, Ray, Whiston, and Woodward. Indeed, it is less impaired than are those other cosmogonies by a desire to account fully for all the minutiae of the Creation and Deluge described in the Bible.

The mysterious Flood

Another system of Earth history, in which the Noachian Flood figures prominently, was projected by the German, Johann Gottlob Lehmann (1719–67). Lehmann's interests spanned chemistry (he died of wounds received from the explosion of a retort filled with arsenic), mineralogy, mining, and geology. From an extensive knowledge of rocks in Prussia, he distinguished in his *Versuch einer Geschichte von Flötz-Gebürgen, betrefend deren Entstehung, Lage, darinne befindliche, Metallen, Mineralien und Fossilien* (published in Berlin in 1756, and regarded by Geikie (1905; 195) as a classic geological paper) three classes of mountains: the first class comprises the primitive mountains which formed coevally with the world and contained no fragments of other rocks; the second class comprises the secondary mountains formed of a succession of well-defined beds which resulted from the partial destruction of the primary rocks; the third class comprises lesser mountains formed by the action of volcanoes and great floods on the secondary mountains. Lehmann attempted to explain the development of the three types of mountain. He believed that the Earth originally consisted of an admixture of earthy matter and water. At the moment of the Creation, the earthy matter was deposited and the water withdrew, some into the central abyss, some to the oceans and lakes. The deposited earth then dried out to form the primitive mountains and valleys. Later, the Noachian Cataclysm occurred—Lehmann believed that the physical cause of this event will ever be a mystery—and the Flood waters overtopped the highest mountains. Earthy material eroded from the primitive mountains by the Flood was held in suspension and then deposited. As the Flood waters retreated, they washed loose earth and animal remains and laid them down as a series of beds in adjacent plains and valleys. This, to Lehmann, explained why the primitive mountains are now bare and have a series of well-bedded deposits along their flanks. Once the material of the secondary mountains had been laid down, they were in some localities acted upon by volcanoes and great floods. Thus was formed the less important third class of mountain.

The Flood as a marine transgression

Another remarkable German geologist, and a contemporary of Lehmann's, was George Christian Füchsel (1722–73), physician to the Prince of Rudolstadt in Thuringia. In his spare time, he wandered in the Thüringer Wald, a region of ancient rocks flanked by Permian and Triassic formations. In 1762 he published in Latin a treatise called 'Historia terrae et maris, ex historia Thuringiae per montium descriptionem' ('A history of the Earth and the sea, based on a history of the mountains of Thuringia'). Eleven years later he published *Entwurf zu der ältesten Erd- und Menschenge-schichte* ('A sketch of the most ancient history of the Earth and Man') (Füchsel 1773). Of relevance to a discussion of cataclysms are Füchsel's views on the geological history of Europe. He supposed that Europe had lain beneath the sea until the formation of the stratum known as the Muschelkalk. Fossil remains of terrestrial plants in these marine deposits attested to the presence of land around the margins of the European sea, in the position of the present oceans. This pre-existing land had gradually been covered by the sea, different parts having subsided successively into subterranean caverns. All the sedimentary strata were originally horizontal, and their present deranged condition can be explained by later upheavals of the ground. Thus Füchsel clearly recognized what would today be called marine regression and transgression.

Floods and convulsive upheavings of the sea floor

A German proponent of revolutions in Earth history was Peter Simon Pallas (1741–1811). Pallas was a naturalist and traveller, who, under the auspices of Tsarina Catherine II of Russia, traversed almost the whole of Asia, and found proof that the Caspian Sea had recently in Earth's history been of much greater extent. In his *Observations sur la formation des montagnes et les changements arrivés au globe, particulièrement de l'Empire Russe* (1771), he suggests that the 'tertiary mountains', which contain the bones of 'the great animals of India', were produced by the most recent revolution of the globe; he writes

These great bones, sometimes scattered, sometimes piled in skeletons and even in hecatombs, studied where they lie, have definitely convinced me of the reality of a deluge over our land—a catastrophe the probability of which I admit I had not conceived before having traversed these shores and seen for myself all that can serve there as proof of this memorable event. An infinite number of these bones, embedded with a mixture of slightly calcined Tellines, bones of fish, glossoptera, ocher-impregnated wood, etc., proves beyond doubt that they have been transported by inundations. But the carcass of a rhinoceros, found with his skin entire and with remnants of tendons, ligaments, and cartilages, in the frozen lands on the banks of the Vilyuy, the best preserved parts of which I have deposited in the Cabinet of the Academy, formed convincing further proof that it must have been a most violent and rapid movement of inundation which long ago transported these cadavers to our frozen climates before corruption had time to destroy their soft parts.

Most natural philosophers who have treated of the physical geography of the world agree in considering all the isles of the South Seas as elevated on immense vaults of a common furnace. The first eruption of these fires, which raised the floor of the very deep sea there and which perhaps in a single stroke or by rapidly succeeding throes gave birth to the Sunda Isles, the Moluccas, and a part of the Philippines and austral lands, must have expelled from all parts a mass of water that surpasses the imagination. This, hurtling against the barrier opposed it on the north by the continuous chains of Asia and Europe, must have caused enormous overturnings and breaches in the lowlands of these continents, . . . and surmounting the lower parts of the chains which form the middle of the continents, . . . have entombed haphazardly the remains of many great animals which were enveloped in the ruin, and formed by successive depositions the tertiary mountains and the alluvions of Siberia. (Quoted in Mather and Mason 1939: 124–5)

New cosmogonies from England and Wales

Belief in the geological importance of the Flood was as rife in eighteenth-century Britain as it was in Europe. A large number of works on the Flood were published, the authors of which were, almost to a man, convinced that most of the Earth's surface features had been created by the Noachian Deluge. Ocean basins, continents, tors, sinkholes, the Norwegian fjords, even volcanoes were held to be part of the Earth's diluvial heritage. Of the works produced during the middle of this period of diluvial studies, two stand out: *An enquiry into the truth and certainty of the Mosaic Deluge* by Patrick Cockburn (1678–1749), published in 1750, and *A treatise on the Deluge* by Alexander Catcott, published in 1761. Both these authors presented cosmogonies couched in terms which were in accord with hexaemeron, more or less, and provided explanations of how enough water was made available to create a universal Deluge.

Explanations of a universal Deluge

Cockburn expounds a cosmogony in which, during the second day of Creation, most of the waters of the Earth subsided into an underground abyss, which formed the great deep mentioned by Moses; some of the waters went aloft to form the clouds; the rest stayed in place to form the oceans. At the Flood, the Earth's surface, on land and beneath the sea, was cracked, probably by earthquakes. Waters emerged from the abyss, and fell from the clouds. All the waters of the Earth were then transported over the land, where they fashioned mountains and valleys. At the Flood's end, most of the waters were returned to the abyss through the still open cracks. Some waters evaporated, but enough were left at the Earth's surface to increase the area of lakes and seas, and to promote the formation of marshes.

Catcott envisioned the original creation of the Earth to involve the formation of a crustal shell, with the shape of a sphere, on both sides of which were waters. Inside the crustal shell, and holding the water against its concave surface, was air. Outside the shell, and holding water against its convex surface, were air and light. The central air expanded, owing to the leaking of light through the crustal shell. The expansion led to the cracking of the crust. Water on the outer surface of the crust then rushed down through the rents and, in so doing, furrowed the smooth land surface into mountains and valleys. The waters moving down met the waters held on the underside of the crust, and they combined to form the waters of the great abyss, a spherical fluid body bounded by the concave surface on the underside of the crust. The shell of the Earth was not utterly destroyed by the retiring waters because their descent was impeded, and their speed was lessened, by air ascending from inside the Earth. Consequently, the waters produced gently undulating hills and dales, rather than jagged precipices and shattered rocks. Thus it was that the antediluvian world was fashioned. When the time came to cause the Flood, all God needed to do was increase the air pressure on the seas, forcing the ocean waters into the abyss. A further application of pressure caused air to enter the abyss, expelling both the waters of the abyss and the waters of the oceans over the land surface. The tremendous forces exerted as the waters were pushed through the crust caused the crustal shell to crack and dissolve. All minerals and metals were dissolved, but organic matters, partly because they were so small and light, stayed intact. The land, and its mineral, metal, and vegetable products, were not thrown into the sea because the land was not dissolved until the Flood waters had risen to their full height and all was calm. The atomized components of minerals and metals, and vegetables and seeds, then settled back to rest in almost the same places that they had resided before the Flood. At the appointed time, God changed the effects of the air so that it pressed solely on the crustal surfaces.

The Earth was formed anew, with fossils left stranded on the mountains, a sign of the universal Deluge.

To demonstrate how torrents in water draining off the land would sculpture the surface, Catcott devised an experiment which provided a miniature replica of the Earth's topography:

I provided a large vessel of glass, had several holes of different sizes bored in the sides about six inches from the bottom, and stopped each with cork: I then filled the vessel with water; and having pulverized beforehand certain portions of the various strata of which the earth consists, as stone, coal, clay, chalk, &c. I permitted these substances to subside one after another through the water, 'till the terrestrial mass reached about two inches above the level of the holes: and the whole settled in regular layers one upon another, just according to the disposition of things in the earth. I then (with the assistance of another) pulled the corks out of each hole as nearly at the same time as possible. The water immediately began to drive the earthy parts through the holes, and scooped or tore the surface of the earthy mass. (Catcott 1761: 170–1)

Collier says that the work of Cockburn, Catcott, and their coadjutors was

a definite attempt to harmonize the Biblical narrative with the accumulations of scientific data that had been so busily gathered during the preceding half-century. It was so far successful as to establish for many decades the orthodoxy of the doctrine that the Noachian deluge was universal rather than limited in extent. That dogma outlived the various hypotheses by which the cause of the deluge and the constitution of the earth were related to the universality of the catastrophe. (Collier 1934: 241)

A universal Flood and a rising sea floor

During the last quarter of the eighteenth century, yet more cosmogonies were proposed. The authors of these cosmogonies— Whitehurst, Williams, and De Luc—all agreed that the Flood had been universal, but they each had their own ingenious explanation of its cause.

John Whitehurst (1713–88), the English clockmaker and geologist, published his theory of the Earth in 1778 in a book with the title *An inquiry into the original state and formation of the Earth*. Whitehurst proposed that the Earth commenced as a chaotic, fluid mass:

Therefore, when the earth was in a state of fluidity, its component parts, solids and fluids, were uniformly blended together, and thus composed one general mass or pulp, of equal consistence and sameness in every part, from its surface to its center. (Whitehurst 1778: 10–11)

Like substances in the chaotic mass attracted one another, and so separated out to form a solid globe of earthy materials, a continuous envelope of water, and an outer envelope of air. Whitehurst agreed with Woodward that the rocks in the solid globe settled according to their relative densities; but, unlike Woodward, he proposed that the precipitation of rocks from the original chaotic mass was, because of the gravitational attraction of the Moon, unequal from place to place. He argued that the attractive power of the Moon would produce tides which would move solids from place to place leading to inequalities in the depth of the sea, and eventually causing dry land to appear. In effect, he envisioned great mounds and hollows on the sea floor produced by tidal action. He explained that the Earth's strata were deposited at this time by the union of similar particles, and were arranged concentrically round the centre of the Earth. Sandbanks built up on the great mounds in the universal ocean and eventually rose above the level of the ocean. These first areas of dry land were man's paradisiacal home. Then fires broke out within the Earth, causing the crust to expand and rise:

Subterraneous fire now being universally generated in the same *stratum* or central part of the earth, by its expansive force gradually distended their incumbent *strata*, like a bladder forceably blown, and, by elevating the bottom of the ocean more than the primitive islands, deluged the whole earth. (Whitehurst 1778: 192)

Whitehurst argued that the weight of sediments lying atop the oceanic mounds surmounted by islands kept crustal uplift in check, so that when the crust expanded, it was the ocean floor which rose. The rise of the ocean floor caused water to wash over, and drown, the primitive islands, an event identified by Whitehurst as the Noachian Flood. The process of emergence was violent:

Subterraneous fire still increasing, its expansive force gradually burst the incumbent *strata*, and opened their fissures more and more, until the two oceans of melted matter and water came into contact, whence a violent explosion ensued, which tore the globe into millions of fragments, and threw them into every possible degree of confusion, some of them being more elevated, and others more depressed. Hence arose an infinite number of subterraneous caverns, apparently many miles, or many hundreds of miles, below the bottom of the primitive ocean. Into these caverns the waters descended, and left the mountains and continents naked and exposed, which had no existence prior to that aera. (Whitehurst 1778: 192–3)

Thus the rise of the ocean floor broke the Earth's crustal strata, permitting seawater to drain into the Earth's interior where, on contact with molten rock, it vaporized in a series of violent explosions. These explosive events fashioned the present continents, which are simply piles of ruined strata. The modern oceans mark 'the watery graves of primitive islands' (Davies 1969: 132). Whitehurst also believed that the cracks in the continental strata, produced during the episode of uplift and explosion, now act as a safety-valve, allowing the slow escape of terrestrial heat which would otherwise accumulate and then escape suddenly and violently, causing a second Flood.

A universal, dissolving Flood

John Williams (1730?–97), a Welsh miner and geologist, explained in his book *The natural history of the mineral kingdom* (1789) that the world was totally destroyed by the Flood, its broken remains being dissolved in the diluvial waters. As the waters abated, the debris was precipitated in a carefully ordered sequence: first granites, then beds of successively younger strata, then, near the top of the sequence, coal measures which were the remains of antediluvian forests. Williams's theory is redolent of Woodward's but differs on one important point: Williams insisted that the precipitation had been uneven, not owing to the differences in gravitational attraction, but to the nature of the diluvial tidal system. Davies explains Williams's reasoning on this matter:

He [Williams] believed that the normal celestial influences had thrown the universal, diluvial menstruum into two antipodal tidal bulges of such size that their passage submerged even the highest of the Earth's present mountains. The movement of the bulges around the Earth, however, was irregular. Because of their momentum, the bulges tended to advance ahead of the solar-lunar pull, so that every twelve hours the bulges became stationary for a time until the celestial influences caught up and allowed the tidal advance to be resumed. This tidal still-stand occurred repeatedly over the same portions of the Earth's surface, and as a result the Afro-Eurasian land complex was built up beneath one bulge, while the Americas were precipitated beneath the antipodal bulge. Between these two stadial positions, the tides ran freely, so that precipitation was at a minimum, and these gaps are today represented by the ocean basins.

Having explained the continents and ocean basins—the first order landforms—as the result of uneven precipitation, Williams went on to offer a similar explanation for such second order features as the world's major mountain chains. These, he claimed, were formed by intense local precipitation, and the highest mountains lie in the tropics because there the tidal bulges were best developed. Similarly, he insisted that mountain ranges, and the 'course or bearing' of strata (the 'strike' in modern terminology), all trend parallel to the axes of the diluvial bulges. These moved fastest in low latitudes, and mountains and strata therefore strike south-eastwards in the southern hemisphere, and north-eastwards in the northern hemisphere. The north-eastward striking Caledonian structures of Scotland and his native Wales had clearly not escaped Williams's notice. (Davies 1969: 139–40)

To explain the disrupted nature of the strata, as seen in present day landscapes, Williams looked to diluvial torrents rather than earthquakes. He suggested that great diluvial tides surged across the young strata every twelve hours as the tidal bulge passed by, cutting long and deep channels in the mountains, and scooping out all the gulfs and deep bays in the oceans.

It is clear from his writing that Williams professed to have little truck with theoretical speculation, preferring to develop ideas out of field investigation:

A detailed account of the speculations of philosophers concerning the original formation of the earth, or of the successive changes to which it has been subjected, might afford some amusement to the reader, and might not perhaps be altogether devoid of instruction, as it would exhibit, in a striking light, the rashness, folly, and presumption of the human mind, in overleaping the bounds of sober investigation and calm enquiry. It is a more difficult task to examine nature herself, to collect and to arrange facts, and to estimate the conclusions which may be fairly deduced from them; than to suppose, with Buffon, that the earth was a spark separated from the body of the sun by the collision of a comet; or with Burnet, that the antediluvian state of the earth gave place to the present, in consequence of the heat of the sun producing fissures and openings, by which the waters of the central abyss burst forth; and part of the surface of the globe having fallen in, now shows the ruins of a former world. (Williams 1810: ii. 559–60)

Although much of Williams's work is based on his own field experience, some of his ideas are as speculative as Buffon's: the empirical pot is calling the theoretical kettle black.

The Flood and crustal collapse

Jean André De Luc (1727–1817), a Swiss scientist and meteorologist, spent half his life in England. He took up residence in England in 1773, and became a tutor to Queen Charlotte and a member of the Royal Society. De Luc thought that the Flood was sufficiently recent an event to have left unmistakable traces, and he deemed it to be the last catastrophic event in the Earth's long history. De Luc's system of Earth history is expounded in a number of his publications dating from 1778 to 1813. Its gist can be gleaned from his five-volume *Lettres physique et morales sur l'histoire de la terre et de l'homme* (1778). The bulk of these books contains detailed descriptions of strata, and betrays De Luc's love of field research:

He was indefatigable in struggling through trackless territories and over cultivated fields, up innumerable mountain peaks and down mines, and in meeting all difficulties including the necessity for communication largely by signs with the illiterate peasantry, in search of facts to prove his theses or to suggest alternatives. (Collier 1934: 264)

De Luc unfolds his cosmogony in the fifth volume of the set. He follows Genesis very closely, though he accepts that the six 'days' of Creation were in fact six periods of great but indefinite duration. He thus divides Earth history into six periods, each period corresponding to one day of the hexaemeron. In the beginning, the Earth, along with all the other heavenly bodies, was a mass of particles—De Luc calls them pulvicles—in a state of rest. On the first day or period, the action of light on the pulvicles set in train chemical processes which produced all the geological phenomena. First, a heavy, turbid liquid formed a sphere with a diameter roughly the same as the present terrestrial diameter. From this liquid, which held the elements of the Earth's present rocks in solution, all the substances of the globe and of the atmosphere separated in succession. At this time, the Earth's core was still only pulvicles. The second day or period saw the precipitation of a bed of slime or mud on the nuclear core of pulvicles, followed by the precipitation of the first of the mineral layers—a very thick crust of granite and similar rocks. In period three, primitive rocks, including gneiss and schist, were precipitated. Inside the Earth, the mud layer which lay beneath the granite crust seeped into the pulvicular core. Some solid parts were formed, which supported the crust, but caverns also formed where the pulvicles had subsided. The result was a series of solid tiers, irregular in shape and resting on one another, which served as a scaffolding. By the end of the third period, the cavern system was extensive. Then, owing to a general subsidence of the pulvicles and caverns, the rock pillars in the cavities gave way, and the weight of overlying deposits caused the crust to sink until all the tiers over large areas of the Earth's surface had collapsed. The sea rushed in, filling the depressed areas, and the level of the primitive ocean sank, exposing the uncollapsed regions as the first continents and islands, on which vegetation appeared. During period four, the Sun and stars became luminous. Chemical precipitation continued in the Earth's ocean basins, many of the rocks deposited being breccias and conglomerates containing the debris produced by the preceding episode of crustal collapse. During the fifth period, the new source of light—the Sun—triggered the precipitation of a new mineralogical type, and the first animals appeared in the seas. The secondary rocks formed during this period thus contain fossils. Another major episode of crustal collapse occurred beneath the oceans leaving submarine mountain ranges, hills, and plains.

The opening of the sixth period saw the precipitation of unconsolidated deposits, including sand and other loose surface strata. The liquid from which all the rocks had by now been precipitated remained as ordinary seawater. The sixth period lasted until the Noachian Flood. During this time, the Earth was quiet, though a slow seepage of liquid under the crust of the continents was opening up a new generation of caverns. Eventually, four thousand years ago, in the time of Noah, the lowermost vaults fell in and the upper levels crashed down on top of them, producing a third episode of crustal collapse, an episode so widespread and convulsive that the positions of the land and sea were completely reversed. As the oceans sank, so the present continents and islands appeared and the Earth's surface assumed its present configuration.

Today, De Luc's views appear as fanciful as the systems of the earlier generation of cosmogonists. The difference between them lies in the manner in which they were derived. Whereas the Restoration cosmogonists relied primarily on the Scriptures, and only to a small extent upon field evidence, De Luc, while not rejecting Genesis, was devoted to studying phenomena in the field and hypothesizing according to the evidence of his eyes. He was no armchair speculator:

> In his truly scientific emphasis on exhaustive accumulations of data and on theories to fit natural phenomena even when the examples contradicted his preliminary hypotheses, as with regard to the inclination of mountainous strata, he is a striking example of the increasing dependence upon observation rather than upon authority which characterized the seventeenth and the eighteenth centuries and which has proved the guide for modern scientists to such comprehension of the world as we have acquired. In another respect he was in line with the best tradition. He broke away from the vulgar abuse of his contemporaries that stained much of the scientific writings of the time as it did the political. Though he criticized, he did so with dignity, justice and kindness, and often expressed appreciation. He never descended to railing, and his objections were leveled against the doctrines rather than against their champions. (Collier 1934: 280–1)

German geognosy

Global floods and bouts of oceanic subsidence

It is ironical that, although the emphasis in geological work during the eighteenth century had shifted from generalization to careful and detailed observation, the first major contribution to the new, empirical science of geology, made by the German mineralogist Abraham Gottlob Werner (1749–1817), involved substantial generalization from limited observations (Hallam 1983: 1). Werner called his new science geognosy, a term coined by Füchsel to define the study of the solid body of the Earth, and the various minerals of which it is composed, as a whole. Werner's system of Earth history was detailed in a private, but widely read, treatise published by a friend of Werner's in 1787 (Adams 1938: 217).

In essence, Werner conceived 'the birth of the world in the bowl of a mighty ocean, each different layer of rock marking a temporary advance of the waters and the laying down of a fresh stratum by the deposition on to the surface below of the heaviest sediments from the massive aqueous solution' (Chorley *et al.* 1964: 25). He believed that there had once existed, at the birth of the Earth, a universal ocean containing in solution all the material that was later to form the Earth's crust. This ocean had intermittently subsided (where it subsided to one is left to surmise) and out of it had precipitated the crustal rocks. The bouts of subsidence in the primitive ocean were, according to Werner, neither slow nor quiet; rather the primitive ocean

was swept and driven by the furious winds and great storms which characterized those times of universal chaos, and also from time to time for reasons that are unknown, it experienced oscillations in level. Powerful and shifting currents set up by the winds and by the draining away of the subsiding waters, cut deep channels through the sediments in all directions and by their erosive power gave rise to deep valleys separated by high mountains. (Adams, 1938: 223)

To Werner then, like Buffon before him, the present form of the Earth was produced by the action in the past of powerful marine and other agencies. Werner envisaged the process of oceanic subsidence and precipitation as a sequence of distinct periods. The first period, or *Urgebirge* (primitive), which Werner had originally termed *Uranfrängliche Gebirge*, saw the chemical crystallization of primitive rocks (granite, gneiss, schist, serpentine, quartz, and so on, with no fossils) during the turmoil of the Earth's birth. The second period, or *Übergangsgebirge* (transitional), which Werner added to his system in 1797 (Adams 1938: 219), saw the laying down of limestones, slates, and shales by chemical precipitation, and of greywackes by mechanical processes. This period, now attributed to the late Palaeozoic, is associated with a lessening of the violent processes of creation, the drawing off of the waters from the primeval ocean, and the deposition of the first organic remains. By the third period, of *Flötzgebirge*, mechanical precipitation had become the dominant process. Limestones, sandstone, gypsum, coal, chalk, and salt deposits were laid down during quiescent intervals when ocean waters again covered the land. Between the quiet episodes occurred bouts of upheaval associated with more violent processes, which produced ore-bearing rocks and basalt. This third period is now attributed to the succession of deposits ranging from the Permian to the Tertiary. The fourth period, or *Aufgeschwemmte Gebirge* (swept together; derivative), saw mechanical precipitation of relatively unconsolidated sand, clay, pebbles, and soapstone in a gradually diminishing ocean. Finally, after a fairly long interval of time, a last bout of violent volcanic outbursts—*Vulkanische Gesteine*—induced by the ignition of underground coal beds, produced layers of lava, ash, and tuff on the land surface.

A Scottish view of geognosy

A system of Earth history similar in principle to Werner's system was promulgated in 1791 by Déodat Guy Sylvain Tancrède Gratet de Dolomieu (1750–1801), after whom the Dolomite Mountains were named. Dolomieu believed that primitive rocks were slowly deposited as horizontal layers in a primeval ocean. A worldwide catastrophe then occurred which, with a force of extreme violence, disturbed the horizontal layers to produce the primitive mountains. There then followed an epoch in which enormous inundations periodically disturbed the deposits.

Further support for Werner's thesis came from the Scottish geologist, Robert Jameson (1774–1854). In the third volume of his *Elements of mineralogy*, entitled *Elements of geognosy* (1808), Jameson brought Werner's ideas to Britain. He was totally won over by Wernerian doctrines and considered all other theories of the Earth useless:

> We should form a very false conception of the Wernerian Geognosy, were we to believe it to have any resemblance to those *monstrosities* known under the name of *Theories of the Earth*. Almost all the compositions are idle speculations, contrived in the closet, and having no kind of resemblance to any thing in nature. (Jameson 1808: 42)

Jameson's theory follows Werner's very closely. Jameson believed that the major relief features of the Earth were produced from the uneven precipitation in the waters which once covered the entire globe, but have since undergone a slow diminution. He argues that the highest mountains and uplands are primitive features, produced by the uneven chemical precipitation of primitive rocks (*Urgebirge*) on the Earth's original solid spherical core (Jameson 1808: 74 n.). The lower mountains and hills are associated with Transitional and Flötz rocks which fill the gaps between the primitive features, and which were chemically precipitated and mechanically deposited when the global waters reached successively lower levels. He does point out that these broad relief features have been modified by other processes, including scouring by currents in the retreating waters of the most recent Flötz transgression and more recent denudation which produced sands, gravels, clays, and alluvium (*Aüfgeschwemmte Gebirge*). Of the most recent Flötz transgression, Jameson writes

It is evident from the nature and position of these rocks, that they have been formed by a vast deluge. The water appears to have risen rapidly; again to have become more calm; and, during the period of its settling, to have deposited the different rocks of this formation; and, lastly, to have retired to its former level with considerable rapidity. (Jameson 1808; 84)

An Irish view of geognosy

Another theory of the Earth, in which the Flood is said to have rushed over the lands from the oceans, was proposed by the Irish lawyer, chemist, and mineralogist, Richard Kirwan (1733–1812). His ideas on this theme are contained in his *Geological essays* (1799) and in a series of papers that he presented to the Royal Irish Academy between 1793 and 1800 (e.g. Kirwan 1793, 1797). Kirwan believed that the rocks of the Earth's crust had been precipitated from a primordial fluid, and that the Earth's topography resulted mainly from the unevenness of the original precipitation. However, he argued that the unevenness of precipitation was not due to Williams's tidal mechanism, but to a random process akin to the precipitation around random local centres as seen in a chemist's retort. He saw primitive mountains as gigantic crystal agglomerations, and plains as areas of minimum precipitation. Once the primitive topography was established, the level of the primordial fluid sank, partly because volcanoes scooped out the ocean basin in the southern hemisphere (how, he does not say), and partly because some of the fluid sank into primitive vaults. The sinking of the fluid led to the emergence of primitive continents which dried out and consolidated. While the fluid was still 9000 feet above its present level, fish were created. The level of the fluid then continued to drop for several centuries, during which time the secondary strata were laid down to form Secondary mountains along the flanks of the Primitive mountains.

To account for the uneven distribution of primitive rock, Kirwan resorted to a tidal theory akin to, if not the same as, that elicited by Williams. He argues that the primordial fluid had been stirred by two sets of tidal movements:

Firstly, there was a normal east to west tidal system resulting from celestial influences, and secondly, there was a vigorous north to south movement caused by 'the water trending to those vast abysses then formed in the vicinity of the south pole'. These two tidal streams, he held, were responsible for giving mountains their widely observed asymmetry, because the tides caused precipitation on the northern and eastern faces of mountains, which were therefore made gentle, while the other faces received no such accretion and remained steep. (Davies 1969: 143–4)

When the fluid had finished retreating into the Earth's cavernous interior, and the Secondary strata had hardened, the globe suffered the cataclysm of the Flood. According to Kirwan, the Flood started in the great southern ocean. He took this view for four reasons: the southern ocean is the largest collection of waters on the face of the globe; the spoils of animals from southern countries and the marine exuviae of southern seas are spattered between latitudes 45° and 55° in the northern hemisphere; traces of a violent shock from the south are perceptible in many countries; and the southwards tapering shape of the southern continents, which suggests that they bore the brunt of the surging Flood waters, only the mountains being able to withstand the violent battering (Kirwan 1799: 68–72). The flood swept northwards, reshaping the continents, giving them their southwards taper and shattering a primitive land mass in the north Pacific region to leave a few islands. In rushing over Asia and North America, the Flood smashed into mountains and scoured the soil leaving barren places such as the Gobi desert. The chief force of the Deluge appears to have been directed northwards between the meridians 110 and 200 degrees east of London. On reaching Siberia, the waters became stationary for a while, but having collected in the Arctic regions, they must have moved southwards. Kirwan argued that the Earth's crust remained unstable for a long time after the Flood, a series of earthquakes associated with the settling of crustal blocks occurring until about 2000 BC. This recent phase of crustal settlement produced, among other features, the Irish Sea, the Straits of Dover, and the Bering Straits.

Changing views of the Flood

It has been established that, during the sixteenth and seventeenth centuries, the Noachian Cataclysm was very popular with students of Earth history. The reason for the popularity of the Flood was, chiefly, that it was described in the Scriptures and so was not an affront to the pervasive bibliolatry of the time. Although the Flood figures prominently in virtually all discussions concerning the Earth and its development, little field evidence was advanced which demonstrated the effect of diluvial waters in fashioning the Earth's surface. During the eighteenth century, the Flood still had a central role in most versions of Earth history, but seeds of doubt appeared as to such matters as whether there had been just one flood. Robert Townson, in his *Philosophy of mineralogy* (1794), pointed to the many changes which have occurred on the Earth as revealed by fieldwork:

our globe, or rather its surface, is not the simultaneous formation of the Omnipotent *fiat* but the work of successive formation and subsequent changes . . . [which are] strong hints, or rather indisputable proofs, of great revolutions. (Townson 1794: 4)

Making the same point, Richard Joseph Sulivan, writing in his six-volume popular epic *A view of nature* (1794), explains

Thus succeed revolution to revolution. When the masses of shells were heaped upon the Alps, then in the bosom of the ocean, there must have been portions of the earth, unquestionably dry and inhabited; vegetable and animal remains prove it; no stratum hitherto discovered, with other strata upon it, but has been, at one time or other, the surface. The sea announces every where its different sojournments; and at least yields conviction that all strata were not formed at the same period. (Sulivan 1794: ii. 169–70)

It being no longer necessary to cram Earth history into the brief chronology of the Bible, geologists at the end of the eighteenth century were more inclined to contemplate the role of sublunary processes in shaping the Earth. The denudation dilemma—explaining how ordinary processes of nature could possibly cause significant change in the Earth's topography in just 6000 years—evaporated, and by the conclusion of the eighteenth century the active role played by denudation was widely recognized, and revealed in three lines of work (Porter 1977: 161–4). Firstly, studies of strata suggested that they were the product, not of a divine fiat at the Creation or Flood, but of natural, sublunary processes. And, in some strata, land and sea fossils were found to occur in alternate beds, suggesting that land and sea changed places several times. Secondly, studies of earthquakes and volcanoes revealed new and convincing evidence that the crust was subject to massive transformations by natural processes. And thirdly, observations of the sea, rain, debacles, and ice in action, made by geologists travelling through Europe and North America, suggested that these natural processes were a force to be reckoned with, quite capable of reducing mountains and cutting valleys.

6

The High Tide of Cataclysmic Diluvialism
Floods and their signatures

By the start of the nineteenth century, the unique role allotted to the Flood in forming the Earth's features had been undermined:

One Deluge—even if a unique event of this kind were still scientifically credible—could not explain the multiple, diverse and complex phenomena. To interpret landforms, it had become necessary to invoke a succession of natural débâcles, and in addition the continuous operation of climate and oceans. (Porter 1977: 163)

None the less, the majority of geologists still believed in the Flood. They did so partly because the bibliolatry of the previous centuries lingered on, albeit in a diluted form, but mainly because, during the first two decades of the nineteenth century, field investigations unearthed a considerable weight of evidence pointing to the existence of a recent cataclysm. The most convincing evidence was the large erratic boulders which lay strewn over much of Britain, Europe, and North America, and the widespread mantle of boulder clay and shelly and bedded drift which was so widespread in the northern hemisphere. When the significance of these deposits was first considered, it was not realized that ice-sheets had once extended well to the south over Europe and North America. In the absence of a glacial theory, it was not unreasonable to think that large boulders resting high in mountains far from their place of origin, and extensive blankets of ill-sorted gravels, sand, silts, and clays, had been deposited by a grand cataclysm. To describe the deposits of mixed and unsorted sediments which lay as a mantle across so much of the northern continents, the term 'diluvium' was coined by Buckland. The term served two purposes: to indicate that the sediments had been laid down during a flood; and to distinguish flood deposits from the 'alluvium' formed by fluvial action. The terms 'alluvium' and 'diluvium' were both quickly adopted by geologists.

Floods in France and Switzerland

The great Swiss debacle

One of the first of the new generation of field geologists with a belief in diluvialism was the Genevan naturalist, Horace-Bénédict de Saussure. In his *Voyages dans les Alpes* (1779–96), de Saussure described the occurrence of erratic boulders of granite in the high passes of the Jura mountains, fifty miles from their source in the central chain of the Alps, and concluded that they had been emplaced by a great debacle, or widespread current of enormous power. The cause of this debacle was, he believed, a geological period when gigantic inthrows of the crust had taken place. The waters of the oceans rushed into the crustal basins, tearing up, fragmenting, and scattering large masses of rock. The Russian geologist, Count Grigory Razumovsky (1759–1837), came to a similar conclusion when he saw enormous erratic boulders in the countryside around Lausanne, Switzerland. He averred that they were carried by a strong torrent of water, their size indicating that it had a power far in excess of the strongest of modern torrents. He proposed that the torrent was produced by a cataclysm of astonishing magnitude. However, he proposed that the cataclysm was the result of natural, rather than supernatural, causes (Razumovsky 1789).

The record in the French rocks

The rocks of France provided a number of late eighteenth-century French geologists and natural historians with some key clues about the nature of the stratigraphic record. The chemist G. F. Rouelle (1703–70) was so struck by the symmetry of the rocks of the Paris Basin, that he used to discourse to his students on it in the Jardins des Plantes. Rouelle never committed his views to print, but Nicholas Desmarest must have heard his talks, and gives a broad outline of his ideas in the first volume of his colossal *Géographie physique* (1794: 409–31). Rouelle thought that the shells embedded in the rocks were significant because they attested to the fact that the rocks were not disposed at random, as had been supposed. He noticed that the shells varied from one region to another, and that certain types of shell tended to occur together, while others were never found together in the same strata. He observed, as Guettard had done before him, that in some districts, the fossil shells were grouped in exactly the same kind of arrangement and distribution as on the floor of the present sea, a fact which, in Rouelle's eyes, disproved the notion that these fossil marine organisms had been brought together by some violent deluge. Rather, it suggested to him that the present land had once been the bed of the sea, and had since been dried out by some revolution that took place without disrupting the strata.

The advanced ideas of Rouelle were matched by the equally enlightened views of Jean Louis Girand, Abbé Soulavie de Nîmes (1752–1813), the French author, churchman, and natural philosopher who may be regarded as the instigator of the field of stratigraphical geology in France. Soulavie describes in his *Histoire naturelle de la France Méridionale* (1780–4) the calcareous rocks of the mountains around Vivarais. He divided the limestones into five ages or epochs, each characterized by a distinct assemblage of fossil shells. The deposits of the last age consisted of conglomerate and modern alluvium, and contained fossil trees as well as bones and teeth of elephants and other animals. He thus realized that the fossils found in rocks could be used to establish an historical chronology.

A theoretical explanation of the nature of the rocks in the Paris Basin was made by Antoine Laurent Lavoisier (1743–94), the famous chemist and victim of the guillotine. Lavoisier published in 1789 a memoir on the horizontal strata of the Paris Basin (see Mather and Mason 1939: 126–8). He distinguished what he called littoral banks and pelagic banks, formed at different distances from the land, and characterized by different organisms and sediment. He thought that the different strata, for instance in the Seine Basin, indicated that the sea level oscillated very slowly, and that a section through all the stratified deposits between the coasts and the mountains would reveal an alternation of littoral and pelagic banks, the number of strata reflecting the number of marine incursions. He thus offered a more sophisticated model of gradual marine transgression than that offered by Aristotle and his followers.

A model of marine transgression was also devised by Jean Baptiste Pierre Antoine de Monet, Chevalier de Lamarck (1744–1829). Lamarck wrote very little on geology, but his *Hydrogéologie*

(1802) contains ideas concerning changes of sea level. Whereas Lavoisier proposed slow changes of sea level, Lamarck envisaged protracted but violent changes. He argued that the Earth has undergone a series of revolutions which are revealed by the fossil record. As to the cause of the revolutions, he accepted that the Noachian Flood or some other great cataclysm may be elicited, providing it is accepted that it can have acted over the vast periods of time required for the deposition of the marine strata. He would, however, seek a cause for the revolutions in the ordinary Earth processes, and not in the supernatural. He argued that the antiquity of the Earth is vast and, given virtually limitless time in which ordinary processes can act, he speculated on the secular westwards displacement of the ocean, and the concomitant submergence of the land. He thought that the ocean slowly moved to the west, eating its way through the land, until it had travelled round the globe, whence the process started anew. During each cycle, of which he thought there were many, the land was reduced by marine agencies to become the ocean floor. New land was created by a shift in the Earth's centre of gravity, produced by the changing position of the ocean basins.

Revolutions in France

It is against a background of lively stratigraphical enquiry in late eighteenth-century France that the work on the Chalk and overlying Tertiary formations of the Paris Basin carried out by Baron Léopold Chrétien Frédéric Dagobert (pseudonym 'Georges') de Cuvier (1769–1832), the celebrated French naturalist and father of comparative anatomy, should be considered. Cuvier was an intellectual giant of his day: 'He had a brilliant analytical mind, a memory like a computer, and the political savvy of a Machiavelli and a Richelieu rolled into one' (Faul and Faul 1983: 137). He was also an out-and-out catastrophist. In 1812, after many years of painstaking research into the geology of the Paris Basin, in which he collaborated with Alexandre Brongniart (1770–1847), he published his masterly work on fossil quadrupeds entitled *Recherches sur les ossemens fossiles* (1812a). In the introduction to this work, which he also published separately in 1812 under the title *Discours sur les révolutions de la surface du globe* (1812b), he proposed that the Earth had suffered not one but many catastrophes in the form

of global earthquakes. To an extent, this view echoed the
suggestion made by the Swiss naturalist, Charles Bonnet (1720–
93), that fossils are the remains of extinct species which had died in
global catastrophes, the last of which was the Noachian Flood
(Bonnet 1779).

Cuvier is adamant that the catastrophes are sudden and violent:

These repeated irruptions and retreats of the sea have neither been slow
nor gradual; most of the catastrophes which have occasioned them have
been sudden; and this is easily proved, especially with regard to the last of
them, the traces of which are most conspicuous. (Cuvier 1822: 15)

The traces referred to are the unputrefied carcasses of large extinct
mammals deep-frozen in northern ice. Cuvier was probably the
first geologist to take these animals as an indication of the
suddenness with which catastrophes strike. The last catastrophe,
he writes, left behind in the northern countries

the carcases of some large quadrupeds which the ice had arrested, and
which are preserved even to the present day with their skin, their hair, and
their flesh. If they had not been frozen as soon as killed they must quickly
have been decomposed by putrefaction. But this eternal frost could not
have taken possession of the regions which these animals inhabited except
by the same cause which destroyed them; this cause, therefore, must have
been as sudden as its effect. (Cuvier 1822: 15–16)

He notes the suddenness with which all the catastrophes, not just
the last, struck, pointing to the evidence found in the rocks to
support this view:

The breaking to pieces and overturnings of the strata, which happened in
former catastrophes, shew plainly enough that they were sudden and
violent like the last; and the heaps of *debris* and rounded pebbles which
are found in various places among the solid strata, demonstrate the vast
force of the motions excited in the mass of waters by these overturnings.
(Cuvier 1822: 16)

Each successive catastrophe changed the landscape and annihilated
almost all the animals and plants then living, a new set of animals
and plants emerging in the aftermath:

Life, therefore, has been often disturbed on this earth by terrible events—
calamities which, at their commencement, have perhaps moved and
overturned to a great depth the entire outer crust of the globe, but which,
since these first commotions, have uniformly acted at less depth and less

generally. Numberless living beings have been the victims of these catastrophes; some have been destroyed by sudden inundations, others have been laid dry in consequence of the bottom of the seas being instantaneously elevated. Their races even have become extinct, and have left no memorial of them except some small fragment which the naturalist can scarcely recognize. (Cuvier 1822: 16–17)

Although Cuvier refers to his *Discours* as a 'Theory of the Earth', he did not intend it to be taken as a cosmogony. Indeed, Cuvier thought the earlier cosmogonical systems too speculative and too ambitious, dealing as they did with events such as the origin of the Earth and changes in the Earth's interior, for which no evidence was left (Hooykaas 1970: 284). No, Cuvier was very much a hard-nosed empiricist who preferred the positive data furnished by observation to fanciful systems and contradictory conjectures concerning the origin of the Earth. He saw in the fossil and stratigraphic record evidence of revolutionary changes heralding the start of new geological and palaeontological epochs. The energy required for the changes was, to him, far greater than the ordinary , slow-acting processes on the Earth's surface caused by weathering, sedimentation, and volcanic eruptions. To Cuvier, these ordinary processes could never, even though they should act over millions of years, produce the disruption and overturning of mountain masses such as the Alps. He thought it futile to seek causes of revolutions and catastrophes, the traces of which are manifest in the Earth's strata, among the powers which now act at the surface of the Earth.

Cuvier's catastrophism was not without its critics. H. S. Boyd, for instance, writing in *The Philosophical Magazine and Journal* for 1817, showed the difficulties of squaring it with the events described in Genesis. But, on the whole, it was very popular, particularly in France. Its last nineteenth-century advocate was Alcide Dessalines d'Orbigny (1802–57), a devoted pupil of Cuvier's, who recognized no fewer than twenty-seven catastrophes in the fossil record (d'Orbigny 1840–89). But it would seem that traditions die hard in France: a hypothesis not so far removed from Cuvier's was proposed by G. Simoens in 1937.

Revolutions in the mountains

Cuvier's ideas were given support by the catastrophic scenario of mountain building envisaged by the French geologist, Jean

Baptiste Armand Louis Léonce Élie de Beaumont (1798–1874). In a paper in the *Philosophical Magazine* for 1831, and later in his *Notice sur les systèmes des montagnes* (1852), Élie de Beaumont argued that, because the Earth slowly and continously cools, as maintained by Buffon, its volume slowly and progressively reduces. The reduction in volume produces the uplift of mountains. Élie de Beaumont was emphatic that, although the cooling process is slow and gradual, the effects it produces, including the uplift of mountains, are violent and sudden. He envisioned long periods of quietness punctuated by short periods of revolution, in which convulsive upheavals of submerged land create gigantic waves which rush over whole continents producing cataclysms on a grand scale.

Élie de Beaumont's views were welcomed by the British diluvialists. Sedgwick applauded the idea that the sudden elevation of mountain chains had been followed by mighty waves rushing over whole regions of the Earth (Sedgwick 1834). Lyell was less than enthusiastic:

But I cannot admit that there are sufficient geological data for inferring such instantaneous upheavings of submerged land as might be capable of causing a flood over a whole continent at once. I may also observe, that the reasoning above alluded to seems to proceed entirely on the assumption that the flood of Noah was brought about by *natural* causes, just as some writers have contended that a volcanic eruption was the instrument employed to destroy Sodom and Gomorrah. If we believe the flood to have been a temporary suspension of the ordinary laws of the natural world, requiring a miraculous intervention of a Divine power, then it is evident that the credibility of such an event cannot be enhanced by any series of inundations, however analogous, of which the geologist may imagine that he has discovered the proofs. (Lyell 1834: iv. 149)

Lyell's reservations notwithstanding, Élie de Beaumont's concept of tectonic revolutions proved fruitful, and was elaborated upon by later European geologists. In France, catastrophist systems of Earth history were still being proposed, and catastrophist views broadcast, almost until the twentieth century, although these systems were more concerned with the nature of mountain building than with diluvial action. L. Frapolli in 1846–7 emphasized the distinction between periods of tranquillity with slow upheavals, and epochs of agitation with sudden upheavals, ruptures, and inundation. Like Élie de Beaumont, he saw processes observed today acting in the past, though he allowed that in early epochs some differences might have occurred owing to different temperatures and a different atmospheric composition. None the less, he could see no reason to conjure up fantastical agents to account for changes in catastrophic periods. Élie de Beaumont's thesis was also championed by his follower, Charles Sainte-Claire Deville (1814–76). In his lectures delivered to the Collège de France in 1875 (Sainte-Claire Deville 1878), Sainte-Claire Deville pointed out that many phenomena in Earth history do not repeat themselves. For example, during the course of time, the atmosphere has lost substances which are harmful to living beings. He also followed his mentor in dividing the effects of geological causes into two groups: slow and continuous effects (sedimentation and the gradual elevation of the continents); and sudden and violent effects (the upheaval of mountains). He believed that as time has progressed, so the slow and continuous effects have been losing their intensity. The most recent supporter of orogenic revolutions associated with a contracting Earth is the British astronomer, R. A. Lyttleton (1982).

The Flood in England

The early nineteenth-century English school of catastrophism was led by the flamboyant geologist and theologian, William Buckland (1784–1856). Rupke (1983: 193) objects to Buckland and his followers being referred to as a school of catastrophists, and their views being set in antithesis to the uniformitarian views of Hutton and Lyell. He claims that, rather, they constituted a school of historical geology. Hooykaas (1970: 291), too, dwells on the similarities, rather than the differences, between Buckland and Lyell. He explains that Buckland and his circle were adamant that the physical causes prevalent today also lay behind the phenomena of the most ancient epochs, and that physical laws which govern slow changes, govern catastrophic changes as well. However, there is a danger of overstating the similarities between the catastrophists and the uniformitarians, particularly prior to 1830: Lyell did not believe in cataclysms and catastrophes; Buckland, Conybeare, and others did, being adamant that the Noachian Flood was an actual event which had left its signature in the landscape. Indeed, it was their firm belief in the Flood which led to Buckland and his coterie being dubbed the diluvialists by Conybeare, to distinguish them from the fluvialists.

Relics of the Flood

Few geologists would take issue with Hallam (1983: 41) or Faul and Faul (1983: 120) when they elect William Buckland as the leader of the diluvialists. In his *Vindiciae geologicae* (1820), Buckland describes a world created under God's guidance by a series of massive and catastrophic upheavals. He is most emphatic that rivers, even 'the most violent torrents', are incapable of forming valleys and basins, and instead looks to the Noachian Flood as a source of mighty erosive power:

Again, the grand fact of *an universal deluge* at no very remote period is proved on grounds so decisive and incontrovertible, that, had we never heard of such an event from Scripture, or any other authority, Geology of itself must have called in the assistance of some such catastrophe, to explain the phenomena of diluvian action which are universally presented to us, and which are unintelligible without recourse to a deluge exerting its ravages at a period not more ancient than that announced in the Book of Genesis. (Buckland 1820: 23–4)

He offers nine points which he sees as proof that the Flood took place and was responsible for all recent features of the landscape and especially the valleys and 'glacial' deposits. Since later in the book the virtues of a diluvialist revival will be extolled, six of Buckland's points are worth quoting in full:

1. The general shape and position of hills and valleys; the former having their sides and surfaces universally modified by the action of violent waters, and presenting often the same alternation of salient and retiring angles that mark the course of a common river. And the latter, in those cases, which are called valleys of denudation, being attended with such phenomena as shew them to owe their existence entirely to excavation under the action of a retiring flood of waters.
2. The almost universal confluence and successive inosculations of minor valleys with each other, and final termination of them all in some main trunk which conducts them to the sea; and the rare interpretation of their courses by transverse barriers producing lakes.
3. The occurrence of detached insulated masses of horizontal strata called *outliers*, at considerable distances from the beds of which they once

evidently formed a continuous part, and from which they have been at a recent period separated by deep and precipitous valleys of denudation.

4. The immense deposits of gravel that occur occasionally on the summits of hills, and almost universally in valleys over the whole world; in situations to which no torrents or rivers such as are now in action could ever have drifted them.

5. The nature of this gravel, being in part composed of the wreck of the neighbouring hills, and partly of fragments and blocks that have been transported from very distant regions.

7. The total impossibility of referring any one of these appearances to the action of ancient or modern rivers, or any other causes, that are now, or appear to have been in action since the last retreat of the diluvian waters. (Buckland 1820: 37–8)

In his book *Reliquiae diluvianae* (1823), Buckland confirmed his agreement with the biblical interpretation of Earth history. The Flood, he wrote, swept away all the quadrupeds, tore up the solid strata of the earth, and reduced the surface to a state of ruin. Everything was explained, in Buckland's eyes, as the direct agency of Creative interference. In this book too, Buckland provides further information on the massive power of the Flood:

An agent thus gigantic appears to have operated universally on the surface of our planet, at the period of the deluge; the spaces then laid bare by the sweeping away of the solid materials that had before filled them, are called valleys of denudation; and the effects we see produced by water in the minor cases I have just mentioned [in a previous paragraph, Buckland listed a few examples of the effects of minor catastrophes including the Val de Bagnes dam burst in Switzerland], by presenting us an example within tangible limits, prepare us to comprehend the mighty and stupendous magnitude of those forces, by which whole strata were swept away, and valleys laid open, and gorges excavated in the more solid portions of the substance of the earth, bearing the same proportion to the overwhelming ocean by which they were produced, that modern ravines on the sides of mountains bear to the torrents which since the retreat of the deluge have created and continued to enlarge them. (Buckland 1824a; 236–7)

Buckland does not claim that all valleys were produced by the Flood:

Though traces of diluvial action are most unequivocally visible over the surface of the whole earth, we must not attribute the origin of all valleys exclusively to that action; in such cases as we have been describing, the simple force of water, acting in mass on the surface of gently inclined and regular strata of chalk and oolite, is sufficient for the effects produced; but in other cases, more especially in mountain districts, (where the greatest disturbances appear generally to have taken place,) the original form in which the strata were deposited, the subsequent convulsions to which they have been exposed, and the fractures, elevations, and subsidences which have affected them, have contributed to produce valleys of various kinds on the surface of the earth, before it was submitted to that last catastrophe of an universal deluge which has finally modified them all. (Buckland 1824*a*: 258)

Also in his *Reliquiae diluvianae*, Buckland reports much of his work on cave deposits. In December 1821, he had visited Kirkdale Cave, Yorkshire, which had been discovered by quarrymen earlier in the same year. The cave contained a large collection of fossil teeth and bones belonging to an assortment of animals: hyena, elephant, rhinoceros, hippopotamus, horse, ox, deer, bear, fox, water-rat, and various birds. He explained this curious assemblage of animal remains as the gradual accumulation of carcasses which had been dragged into the cave by hyenas, and had then been covered by a layer of mud washed in by the waters of the Flood. This hyena den theory became enormously popular, despite the fact that it was at odds with the diluvial theory. Traditional diluvialists would have argued that, many of the animals in the cave being tropical species, they must have been transported in the Flood waters from Africa and swept into the cave. But to Buckland, the bones belonged to animals which had lived, died, and been devoured in antediluvian Yorkshire.

Although the hyena den theory seemed to suggest that the Flood was not as powerful as most diluvialists would maintain, Buckland none the less still believed in it. He carried out much fieldwork to establish the origin of gravel deposits and valleys. In a paper on the excavation of valleys by diluvial action, Buckland furnished several examples of valleys in Devon and Dorset that had been cut by the direct action of diluvial currents, and in the same paper he argues that 'glacial' erratics found in southern England had been transported there by diluvial currents from north-east England and the Midlands (Buckland 1824*b*). In a later paper on the formation of the Vale of Kingsclere, Hampshire, he reaffirmed the role of the Flood in fashioning valleys, but acknowledged too that geological structure also affects the form of a valley. He argued that the anticlinal Vale of Kingsclere was not formed by diluvial denudation alone, but by a combination of factors including the local uplift of the chalk to produce a 'valley of elevation' which was then modified by diluvial action (Buckland 1829).

The nature of the Flood

Buckland's views caused a stir in geological and theological circles. Biblical literalists raised scores of objections to Buckland's hyena den theory and his diluted brand of diluvialism, which they voiced in a stream of articles and books (see Rupke 1983: 42–50). Foremost among the geological critics was Lyell. In his *Principles*, Lyell attempts to demolish Buckland's thesis by falling back on the old argument over whether the waters of the Flood came as a violent and transient rush, as Buckland held, or as a quiet effusion of waters upon the Earth, as recorded by Moses. Lyell rests his case by joining John Fleming (1826) and citing the rather dubious evidence of the olive-branch brought back by the dove, which 'seems as clear an indication to us that the vegetation was not destroyed, as it was then to Noah that the dry land was about to appear' (Lyell 1834: iv. 149). The same argument, put in more flowery language, was offered by an anonymous correspondent of the *Philosophical Magazine and Journal* for 1820, identifying himself as A.B.C., who was prompted to write a letter after having perused Buckland's inaugural lecture delivered in May 1819 at Oxford. The comments of this critic are typical of a host of writers who, probably to the pious Buckland's surprise, felt that liberties were being taken with Holy Writ. A few quotations should suffice to gain the tenor of the arguments against Buckland's thesis:

I remember having seen Mr. Bakewell commended in your work, for having in the year 1813 abstained, from introducing the Deluge of Moses into his 'Introduction to Geology,' as the previous writers had almost invariably done, to the manifest injury of geology on the one hand, and of religion on the other: since which, the practice has almost entirely grown into disuse, while the number of writers on geological subjects, have been greatly on the increase; and I regret therefore to see, the new Geological Professor at Oxford, attempting now to revive the exploded notion, that any of the phaenomena at this time *visible*, on or within the Earth, are, with any proper regard to probability, referable to the Deluge of which Moses writes. (Anon. 1820: 10)

This criticism seems not unfair, given that late eighteenth-century geologists had demonstrated the plurality of cataclysms. But Buckland simply believed that he had found evidence for a recent cataclysm which might have been the Deluge described in Genesis; and by the mid-1830s he openly accepted the repeated revolutions and convulsions that have affected the surface of the Earth (Buckland 1836; 547). The critic accepts that the signature of an overwhelming torrent, or perhaps succession of torrents, is found almost universally across the globe. He accepts that the waters of these torrents could transport vast amounts of earthy matters, gravel, and large boulders, lodging them on the tops of hills, and on plains, at very great elevations. But he does not accept that the Earth's topography was fashioned by them:

hills and plains, and the valleys which intersect them, having most evidently existed in their present form and shape, at the time of these early *gravel floods*, which most evidently did not excavate the valleys, or in any material degree abrade or alter the contours of the hills. (Anon. 1820: 11)

This is a particular bone of contention that the critic picks with Buckland. He is adamant that Buckland is wrong in asserting that the Flood waters were violent:

Now the mistake of Professor Buckland, and of all those who have preceded him, in referring *these tumultuous events*, to the Deluge happening in the days of Noah, consists in not having carefully considered *the words used by Moses in describing the Noachian Deluge*, which if they had done, instead of taking on trust, the absurd interpretations of those words, or rather the fabrications which were framed by Dr. Woodward and many other writers of the last two centuries, the Professor must, by this *examination of Moses' words*, have found, that the same, throughout, refer to *a quiet effusion of water upon the surface of the Earth*, for the avowed purpose and for no other, but that of *drowning the degenerate race of Mankind*, whose crimes and violences had filled the Earth; and that in point of fact, *according to Moses, the surface* of the Earth, was not torn up or moved, so as in any material degree to disturb and root up *the vegetable races!*; nor did it annihilate any of *the race of fishes*, not even the most torpid and helpless of the species of shellfish! The vegetable earth or mould, fit for the growth of useful plants (the evidently *slow result* of long periods of decomposition, and the accumulation of decayed vegetable matters) was not, *according to Moses*, either washed away, or covered, by naked and *fresh-moved rubbish*, because Noah on quitting the Ark, or very soon after, *planted a vineyard!* (Anon. 1820: 11)

However, whether the Flood waters were calm or violent, Buckland had found apparently incontrovertible evidence of the Flood in the extensive blanket of diluvium and the erratic blocks found over large parts of the northern hemisphere. Buckland's views on the origin of diluvium from a grand cataclysm were apparently confirmed when fragments of sea shells were discovered in diluvium on Moel Tryfan, near Caernarvon, by Joshua Trimmer (1834), around Dublin (on the promontory of Howth, on Bray Head, and in the valley of Glenismaule) by John Scouler (1836–7), and at other places in the British Isles. Sceptics questioned the lack of bedding and sorting in the diluvium, which consisted of an unstratified mixture of sand, pebbles, and huge boulders. This apparent flaw in the diluvial theory was countered by arguing that, during the grand cataclysm, icebergs had been swept south from polar regions to warmer latitudes where they melted, depositing their cargo of morainic debris. The process of ice-rafting also offered a neat explanation for the strewing of large individual erratic blocks, a problem which had been recognized in the closing years of the eighteenth century by de Saussure.

Causes of the Flood

The English diluvialists, and Buckland in particular, evinced in their early works a belief in one recent grand cataclysm. Buckland would not be drawn on the question of the cause of this deluge, which was perhaps wise, for 'if we would investigate the means by which this tremendous catastrophe was produced, the mind is easily bewildered in unprofitable conjecture' (Greenough 1819: 189). But he clearly believed that it was one and the same event as the Noachian Flood, and visualized it as a giant surge or tidal wave (Rupke 1983: 40). This belief bestows on Buckland the doubtful honour of being the last British geologist of note to reconcile geological discoveries with the Scriptures (Davies 1969: 251). Interestingly, Buckland's hyena den theory undermined the notion, held by such diluvialists as de Luc and Cuvier, that the Flood had been produced by the sudden interchange of land and sea. Clearly, if Buckland's theory was correct then hyenas and their prey had lived on land which had been dry both before and after the Flood (Rupke 1983: 40).

Lyell trod very carefully when commenting on the cause of the Flood, lest he should step on the toes of the pious public of late Georgian England. He simply stated that he had always considered the Flood, when its universality in the strictest sense was insisted upon (that is, when it was deemed global), as a 'preternatural event far beyond the reach of philosophical enquiry, whether as to the causes employed to produce it, or the effects most likely to result from it' (Lyell 1834: iv. 149–50). Other geologists, more adventurous if less prudent perhaps, were prepared to venture guesses as to what might have caused the Flood. George Bellas Greenough discusses the matter in some detail. He recognizes that the cause being looked for must be sought outside the Earth and outside the Solar System, and must be capable of

inundating continents, and giving to the waters of the deep unexampled impetuosity, but without altering the interior constitution of the earth, or deranging the sister planets; moreover the cause must be transitory, and one which, having acted its part once, may not have had occasion to repeat it in the long period of five thousand years. Any supposeable cause that would not fulfil these conditions, is insufficient for our purpose. (Greenough 1819: 196)

With seeming reluctance, he considers, as, without hesitation, Whiston, Halley, Holbach, and others had before him (see Chapter 9), if a comet might not fulfil the conditions:

Much would depend on its bulk and distance. It would not fulfil them if we suppose a comet, large in comparison of the earth, to move in a line joining the centres of the two bodies, so as to produce a direct shock; but, if we suppose one of suitable dimensions to move in such a direction as would allow it only to graze the earth, it is not impossible that the shock of this body, a body, such as we require, out of the solar system, might produce the degree and kind of derangement which we are attempting to account for; I mean a great temporary derangement on the surface of the earth, unaccompanied by any material change of its planetary motion. Euler, who, in a treatise entitled '*De periculo a nimiâ cometae appropinquatione metuendo*,' has investigated the changes that would be made in the elements of the earth's orbit by a comet, its equal in bulk, coming almost in contact with it, finds that the attraction of such a comet would indeed alter the length of our year, but only by the addition of seven hours. The maximum effect resulting from the comet's attraction at the time of its passage, would be greater than we should be led to infer from the total result of its attraction, after its final departure; for the changes occasioned during its approach, would be in great measure undone during its retreat; but even at their maximum they would not be very great, because from the rapidity of the comet's motion, time would be wanting to complete them. A comet grazing the earth, would be incompetent, Euler says, to produce even a deluge of our continents, unless the shortness of its stay were compensated by a magnitude of volume, exceeding that upon which he has founded his calculation. (Greenough 1819: 196–8)

He concludes by remarking that

if the hypothesis of a shock derived from the passage of either a comet or one of those numerous, important, and long neglected bodies, often of great magnitude and velocity, which occasion meteors, and shower down stones upon the earth, would explain the phenomena of the deluge, (a point upon which I forbear to give an opinion,) we need not be deterred from embracing that hypothesis under an apprehension that there is in it anything extravagant or absurd. In the limited period of a few centuries, there is little probability of the interference of two bodies so small in comparison with the immensity of space; but the number of these bodies is extremely great, and it is therefore by no means improbable, says La Place, that such interference should take place in a vast number of years. (Greenough 1819: 198–9)

In 1834, Greenough had second thoughts about the single Flood hypothesis, and recanted his views on the subject, accepting instead that there had been several successive inundations. He was also less warm about the cometary encounter hypothesis. John Henslow, in a paper in *The Annals of Philosophy* for 1823, had already rejected Greenough's suggestion that a close encounter with a comet might have caused a universal deluge on the grounds that the putative effects are incompatible with the appearances observable in the diluvium, and that a comet grazing the Earth would have far more catastrophic consequences than a transient, if universal, flood of waters. Henslow then proposed a new hypothesis in which it is assumed that the continents can act like a sponge, soaking up the excess water added to the Earth during the Flood. He assumed that water was somehow added to the Earth at the time of the Deluge, and that that water is still with us. With the aid of a diagram (Figure 6.1), Henslow expounded his idea:

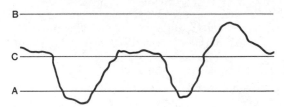

FIG. 6.1. Henslow's diagram showing sea level before (A), during (B), and after (C) the Flood. (From Henslow 1823)

Suppose the original level of the surface of the ocean to have been the line A, and an increase of waters to have raised the surface from A to B, sufficient to cover the tops of the highest mountains; I would ask whether, if the increase were rather sudden as it is stated to have been, we may not imagine that a considerable depression below the highest level would afterwards take place, owing to those solid portions of the earth which were not originally covered, becoming saturated with moisture; and thus, after a certain lapse of time, the surface of the ocean might rest at C, leaving the higher summits of the old continents again exposed. (Henslow 1823: 345)

A very popular cause to which a cataclysm could be attributed was a convulsive movement of the Earth's crust. Adherents to this cause of floods became known as the cataclysmic structuralists. They stressed the dominance of violent movements of the crust in creating the major features of the Earth's surface, with secondary features resulting from the rushing cataclysms of water impelled by the upheavals and guided by rifts and Earth fractures (Chorley *et al.* 1964). The promulgator of this idea was the Cambridge mathematician, William Hopkins. Building on the work of Élie de Beaumont, Hopkins (1838, 1842) suggested that the tremendous jerks associated with the sudden and violent upheaving of the sea floor would cause commotion of an uncommon order in the overlying ocean. Tidal waves would be generated which, because they were set in motion by earth movements or volcanic activity, would not act like ordinary waves of oscillation. In an ordinary wave, suspended particles bob up and down, rather than moving forwards. The tidal waves produced by earth movements would, however, carry suspended particles, including large blocks of rock, long distances. For this reason, Hopkins called these tidal waves 'waves of translation'.

William Whewell (1847) used Hopkins's theory of waves of translation to explain the distribution of the Northern Drift (diluvium), which forms an extensive blanket over northern Europe and Russia. He carefully calculated the magnitude and frequency of upheavings required to produce waves capable of carrying and depositing the Northern Drift. He started by assuming that the drift occupied a semicircle with an 800-mile radius. He supposed that all the drift had been derived from one centre, took an inner circle of 200-mile radius, and considered the drift to occupy the annular space between the two circles. The mean distance from the centre of the annulus was thus 500 miles. Whewell assumed that at this mean distance every square mile is evenly covered with drift some one thousandth of a foot thick, which is equivalent to supposing that there is a patch of drift one-tenth of a mile square and one foot deep on each square mile. He also assumed that the drift has a specific gravity about three times that of water, so the effective specific gravity of the drift when immersed in water would be about twice that of water. He then explains that

The horizontal force which it requires to move a body along a surface on which it rests, depends on the form of the body, its texture, and that of the surface and other circumstances: but I think we may suppose it would require a force and pressure of at least one-fourth the weight of the mass moved, to propel rocks and loose materials along the bottom of the sea.

This being assumed, it will require a force (pressure) equal to the weight of half a cubic foot of water to move a cubic foot of drift; and so, for other quantities. . . .

Now this mass of drift, which is found on an average mile at the mean distance, has travelled 500 miles from the centre. And the labouring force which has carried it through this space, in whatever way it has acted, must be equivalent to the product of the moving pressure and the space through which it has acted; that is, it must be equivalent to the weight of $\frac{1}{1,000,000}$ of a mile of water, multiplied by 500 miles. . . . That is, one cubic mile of water rising through $\frac{1}{2000}$ of a mile (or about 2½ feet) would supply the power necessary to carry the drift which occupies one average mile at the mean distance from the centre of distribution. . . . But we may put this result in a shape more readily conceivable. It is equivalent to 4500 cubic miles of water raised through a space of $\frac{1}{10}$ of a mile; or again, a body of water 45,000 miles in surface and $\frac{1}{10}$ of a mile deep, raised through $\frac{1}{10}$ of mile. If then we suppose a sea-bottom 450 miles long and 100 miles broad, which is $\frac{1}{10}$ tenth of a mile below the surface of the water, to be raised to the surface by paroxysmal action, we shall have the force which we require for the distribution of the northern drift, on the numerical assumptions which have been made. And this is true, whether we suppose the elevation to have taken place at once, or by repeated operations, so long as they are paroxysmal. We shall have the requisite force, for instance, if we suppose this area to be elevated by ten jerks of 50 feet each, fifty jerks of 10 feet each, or by the same 500 feet any how divided into sudden movements. And as we diminish the area elevated, we must increase the total amount of elevation in the same proportion, so as to retain the same ultimate product of water paroxysmally elevated through a certain space. In all these cases, we shall have a machinery, which, operating through waves of translation, will produce the requisite effect. (Whewell 1847: 229–31)

Whewell favoured a train of waves generated by a succession of sudden upheavings, rather than a single wave, because he felt that one wave would be incapable of carrying the load or depositing it over a wide enough area. He also demonstrated that gradual and gentle elevations would not generate waves of translation—only paroxysmal uplift would do that.

Armed with Whewell's jerk magnitude and frequency calculations, Hopkins (1848) set about explaining the topography of the English Lake District. A flaw in Hopkins's theory seemed to be that, if the waves of translation had indeed transported all the drift and boulders to their present locations, then they must have carried sediments up and over relatively high ground. Hopkins was undeterred when this problem was pointed out:

The difficulty in this theory arising from the presumed inequalities of the surface over which the blocks must have been transported, has been, I conceive, in many instances, far too much insisted on; for it has been made to rest on the assumption that the inequalities of surface between the present and original sites of erratic blocks were the same, or nearly so, at the time of transport as at present; an assumption which I regard as totally untenable. There are three obvious causes of inequality of surface—elevation and disruption, denudation during gradual emergence from beneath the ocean, and erosion after emergence. . . . such great inequalities as those presented by the oolitic and chalk escarpments, have doubtless been due in a great measure to denudation, during the period of gradual emergence of the land, the higher levels being raised above the sphere of denuding action, while the lower levels remained exposed to it. (Hopkins 1884: 237)

Hopkins was convinced that his wave of translation theory could account for the spread of drift material, and therefore declined to accept the glacial theory. As far as he was concerned,

A series of uplifts of the order of 100 or 150 feet could produce waves capable of moving drift and erratics. Final uplift would allow wave erosion and current action to operate along faults to produce the present topographic inequalities. As the sea retreated from the rising landmass and a fault was progressively exposed, so the waves would excavate a valley from the head down towards its eventual mouth. (Chorley *et al.* 1964: 344)

One Flood or many?

Faith in a single, universal deluge was shaken during the 1820s and 1830s by two geological discoveries: the lack of a mantle of diluvium in low and middle latitudes, and the complex stratigraphy of the diluvium itself (Davies 1969: 251). To account for these findings, especially the second one, a new diluvialism was created in which multiple diluvial episodes were proposed. The notion of a series of deluges had been promulgated by James Hall in 1812. In two papers read to the Royal Society of Edinburgh, Hall suggested continents emerge not gradually, as Hutton had maintained, but in a series of sudden and violent jerks. Each sudden bout of continental uplift produces a giant tsunami which rushes over the land. Hall's diluvialism was, at the time he proposed it, advanced in that it envisaged repeated, and in some cases local, deluges. Interestingly, when Buckland proposed his single flood hypothesis some seven years after Hall's papers appeared, he took a backwards step by attempting to revive an outmoded idea which harked bark to the thesis of Catcott. It is not surprising, therefore, that Buckland's diluvialism was relatively quickly ousted by a new diluvialism which embraced the notion of recurrent cataclysms as proposed by Hall. As a theoretical backdrop to the views of the new diluvialists, Élie de Beaumont's popular theory of repeated mountain-building revolutions served very well indeed. It seemed that each such revolution must have sent gigantic tsunami surging around the world, inundating and remodelling the continents (Davies 1969: 252). The remodelling of topography was thought to have been achieved in three ways: by the violent impact of the diluvial waters; by the erosive work of diluvial currents during inundation; and by the action of diluvial torrents draining off the continents as the flood waters abated. Most new diluvialists regarded the diluvial torrents as the most effective of the three agents of erosion. In this, they followed the much earlier proposals of Alexander Catcott (1761) and John Williams (1789). Buckland was converted to the new diluvialism in the late 1820s, and, by the time he wrote the Bridgewater Treatise on *Geology and mineralogy considered with reference to natural theology* in 1836, he admitted the folly of ascribing the formation of all stratified rocks to the effects of the Mosaic Deluge (Buckland 1836: 16).

The members of Buckland's coterie were strongly supportive of the Buckland brand of catastrophic diluvialism, though the evidence for multiple cataclysms was too overwhelming for them to stand by the notion of a single deluge. Adam Sedgwick (1785–1873), Woodwardian Professor of Geology at Cambridge, in a paper on the origin of alluvial and diluvial formations (Sedgwick 1825), expressed views very close to those of Buckland, but was inclined to Élie de Beaumont's hypothesis of the sudden and repeated elevation of mountain chains, each successive convulsive uplift causing a cataclysm (Sedgwick 1834). He differed from Buckland only in having some reservations about the connection between the findings of geology and the truths of the Bible: he believed that the two did not contradict one another, but that the methods and modes of dealing with each are very different and they are best left unconfused (Chorley *et al.* 1964: 119).

William Daniel Conybeare (1787–1857), another contemporary of Buckland's, held opinions akin to those held by Sedgwick, though Conybeare's opinions were more firmly based on field evidence, and more coherently expressed. In a paper 'On the valley of the Thames', read to the Geological Society of London in February 1829, Conybeare invokes several cataclysmic agents and a series of catastrophes to account for features he had observed in the field:

He proceeds to distinguish several different geological epochs, at which it is probable that currents must have taken place calculated to excavate and modify the existing surface. I. In the ocean, beneath which the strata were originally deposited. II. During the retreat of that ocean. III. At the periods of more violent disturbance, which are evidenced by the occurrence of fragmentary rocks, the result of violent agitations in the waters of the then existing ocean propagated from the shocks attendant on the elevation and dislocation of the strata.—Four such periods are enumerated as having left distinct traces in the English strata. 1. That which has formed the pudding-stone of the old-red-sandstone, ascribed to the elevation of the transition rocks. 2. That which has formed the conglomerates of the new-red-sandstone, ascribed to the elevation of the carboniferous rocks. 3. That which has formed the gravel beds of the plastic clay. 4. That which has produced the superficial gravel, spread alike over the most recent and oldest rocks as a general covering, and which is found to contain bones of extinct mammalia: this (it is agreed) may be identified as the product of one aera, by the same evidence which is employed to demonstrate the unity of any other geological formation. (Conybeare 1834: 145–6)

Chorley *et al.* (1964: 120) note that, like other diluvialists, Conybeare was misled into thinking that all gravels were indicative of Flood movement, and that conglomerates were deposited at a time of turmoil. Although this is true in that many gravels attributed to the Flood by the diluvialists are in fact glacial in origin, it is also true that many gravels in England, and in southern England in particular, are fluvial or marine in origin, the exact provenance of some of them remaining something of a mystery.

Conybeare looks in detail at the valleys of the Thames Basin and makes a number of points about them: firstly, that their depth and shape suggest that a violent force was involved in their fashioning; secondly, that they cannot have been formed by forces of erosion which operate at present, forces which have left the earthen ramparts of British and Roman camps virtually intact; thirdly, that the wealth of flood deposits in the valleys, lying above the present day flood waters, can only be explained as the work of present rivers if those rivers had continually changed their course, but that the survival of historical sites attests to the permanence of the present river channels; and, lastly, that the cutting of the Thames river through the Oxford hills can only be explained by the action of retreating diluvial waters. In support of the last point, Conybeare cites the evidence of dry valleys:

The Chilterns, like most other chalky districts, abound with dry valleys, the rifted and absorbent structure of that rock not permitting the rain waters to collect into streams: these valleys agree in every other feature with those containing water courses, and have been obviously excavated by the same denuding causes, which, in this case, it is self-evident could not have been river waters. The surface of the chalk has been deeply and violently eroded, and it is deeply covered with its own debris;—this action appears, in part, to have taken place during the epoch of plastic clay formation. (Conybeare 1834: 148)

In an article published in the *Philosophical Magazine* in 1830–1 Conybeare reiterates his views on diluvialism, explaining that catastrophes are caused, like slow processes, by the action of water and volcanic power, but then asserting, somewhat in contradistinction to his first statement, that the excavation of many valleys has been produced by the violent action of mighty diluvial currents and not by the drainage of rainwater through the present river system.

The new diluvialism was energetically advocated by the ardent catastrophist, Robert Impey Murchison (1792–1871). In the fourth and last edition of his *Siluria* (1867), he talked, following the lines set out by Élie de Beaumont, of ordinary operations of accumulation being continued tranquilly during very lengthened periods but such periods being broken in upon by great convulsions. He also had a liking for Hopkins's theory of waves of translation, and argued forcefully in favour of orogeny and marine erosion as major processes in Earth history. In the early 1860s, when the word had almost fallen into disuse by geologists, he was 'still holding aloft the tattered banner of catastrophism', expecting geological opinion to shift round to his point of view (Davies 1969: 346). Even as late as 1869, he was trying to sell to the Fellows of the Royal Geographical Society the view that the cliffs on either side of the Straits of Dover and the Irish Sea had been torn asunder by a violent convulsion (Murchison 1869). He was the last geologist of renown who spoke out for the catastrophist cause in nineteenth-century England. A few other, less illustrious geologists doggedly followed catastrophist doctrines. From a survey of the structural geology of south-east England, Searles V. Wood concluded that the entire landscape had formerly been mantled with thick deposits of Oolitic, Cretaceous, and Tertiary strata, on which the present valley system evolved during two tectonic revolutions (Wood 1864). And, as late as the turn of the century, Sir Henry Hoyle Howorth was flying the threadbare banner of catastrophic diluvialism in the face of the geological establishment, making a desperate plea for a return to the geology of the old catastrophists in his elephantine and assiduously researched works *The mammoth and the Flood: An attempt to confront the theory of uniformity with the facts of recent geology* (1887), *The glacial nightmare and the Flood: A second appeal to common sense from the extravagance of some recent geology* (1893), and *Ice or water: Another appeal to induction from the scholastic methods of modern geology* (1905).

The Flood in America

The father of American geomorphology was Lewis Evans, a noted surveyor, draughtsman, and maker of maps. Born in 1700 in Pwllheli, Wales, Evans emigrated to Pennsylvania. His extensive surveys of the north-eastern United States were reported in a journal—written in 1743, but not printed until thirty-three years later (Pownall 1776)—several annotated maps, and a pamphlet entitled 'An analysis of a general map of the Middle British Colonies' (Evans 1755; White 1956). Evans, like earlier observers of nature in Europe, had seen shells on mountain tops. But, rather than invoking a universal deluge to explain their presence, he resorted to an argument which betrays a shadowy grasp of the concept of isostacy. To explain how the land and sea may change places, he argued that if the ridges of rock impounding an immense body of water in the Great Lakes were to break by a natural accident, or were to be breached by the prolonged cutting of a passage, then the lakes would drain. The Great Lakes region of America, having its watery burden removed, would then rise owing to the immediate shifting of the Earth's centre of gravity, at once or by degrees (see Pownall 1776: 30). Evans also believed in the Noachian Flood. He contended that the Appalachian Mountain region showed evidence of the Flood. The loftier peaks had suffered least from the cataclysm. They stood now at the same height as before the Flood; all they had lost was much of their soil cover. The lower hills and valleys had borne the brunt of the Flood waters, and were extensively eroded (cf. Chorley *et al.* 1964: 237–8).

North American diluvium and erratics

Erratic boulders and diluvium in North America were first described in 1753 by the Scandinavian scientist, Peter Kalm (see

Kalm 1937 edn.). As in Europe, these deposits were seen as firm evidence of a cataclysm. Thus, on discovering sixty-four varieties of rock deposited along the shores of Lake Superior, Benjamin de Witt was prompted to write to the Philadelphia Academy in 1793:

Now, it is almost impossible to believe that so great a variety of stones should be naturally formed in one place and of the same species of earth. They must, therefore, have been conveyed there by some extraordinary means. I am inclined to believe that this may have been affected by some mighty convulsion of nature, such as an earthquake or eruption; and perhaps this vast lake may be considered as one of those great fountains of the deep which were broken up when our earth was deluged with water, thereby producing that confusion and disorder in the composition of its surface which evidently seems to exist. (Quoted in Merrill 1924: 12)

Amos Eaton (1776–1842), a student of Benjamin Silliman and a strongly religious man, was a believer in the Flood as God's instrument of punishment for a wicked world:

The written history of the deluge might be varied more or less by erroneous copies and incorrect translations. But the geological records of divine wrath poured out upon the rebellious inhabitants of the earth at that awful period can never be effaced or changed. (Quoted in Merrill 1924: 131)

He did, however, question the manner in which the Flood could have transported granite boulders in the Connecticut valley:

What force can have brought these masses from the western hills, across a deep valley seven hundred feet lower than their present situation? Are we not compelled to say that this valley was once filled up so as to make a gradual descent from the Chesterfield range of granite, syenite, etc., to the top of Mount Tom? Then it would be easy to conceive their being rolled down to the top of the greenstone where we now find them. (Quoted in Merrill 1924: 617)

Not that the suggestion of filling up the existing valley to allow boulders a gradual descent to their present locations was new; it had been considered, but rejected, by Conybeare as a possible explanation of the gravel deposits found in the Thames Basin (Conybeare 1834: 148–9).

The Flood and melting ice-caps

Borrowing ideas from European diluvial theorists and applying them in the North American setting was standard practice for American geologists in the first half of the nineteenth century. The first volumes of *Silliman's Journal*, later called the *American Journal of Science*, are dominated by papers from American diluvialists. Horace H. Hayden, for instance, was a Baltimore dentist and one-time architect whose profession did not deter him from having firm ideas on geology, and especially the Flood (Chorley *et al*. 1964: 243–4). Unlike his illustrious European counterparts, Hayden declined to believe that the Flood waters had issued from the abyss or had descended from the tail of a comet. He favoured the sudden melting of the polar ice-caps, caused by a change in the tilt of the Earth's rotation axis, as the source of all the Flood water. All alluvium, he explained, was deposited by the Flood waters, and erratics were transported in rafts of ice then dropped at their present locations. Of the unconsolidated Tertiary and Recent deposits of the Atlantic coastal plain, he has this to say:

Viewing the subject in all its bearings, there is no circumstance that affords so strong evidence of the cause of the formation of this plane as that of its having been deposited by a general current which, at some unknown period, flowed impetuously across the whole continent of America, in a northeast and southwest direction, its course being dependent upon that general current of the Atlantic Ocean, the waters of which were assumed to have risen to such a height that it overran its limits and spread desolation on its ancient shores. (Quoted in Merrill 1924: 618)

Hayden's ideas, as expressed in his *Geological essays* (1820), seem to have been taken seriously. In a lengthy review, Silliman agrees that the location of the unconsolidated deposits, far removed from their places of origin, can best be explained by the action of one or more large floods. But although Silliman's thinking is advanced enough to attribute the deposition of the unconsolidated deposits to more than one flood, he is still captivated by medieval and mystic fantasies, with ideas of caverns inside the Earth filled with water which, at God's command, gush out over the surface of the globe (Chorley *et al*. 1964: 246):

The existence of enormous caverns in the bowels of the earth, (so often imagined by authors,) appears to be no very extravagant assumption. It is true it cannot be proved, but in a sphere of eight thousand miles in diameter, it would appear in no way extraordinary, that many cavities might exist, which collectively, or even singly, might well contain much more than all our oceans, seas, and other superficial waters, none of which are probably more than a few miles in depth. If these cavities communicate in any manner with the oceans, and are (as if they exist at all, they probably are,) filled with water, there exist, we conceive, agents very competent to expel the water of these cavities, and thus to deluge, at any time, the dry land. (Silliman 1821: 51–2)

It is very possible, that anterior to the deluge of Noah, and to the peopling of the globe by rational beings, and during the gradual draining of the earth from the grand chaotic deluge, several floods more or less partial or extensive, may have taken place,—thus accounting for partial formations, as the parasitical trap rocks, &c. (Silliman 1821: 53)

More signs of the Flood in North America

A passion for diluvialism was evinced by virtually all American geologists in the first half of the nineteenth century. One reason for this was 'the presence of the Great Lakes, particularly as for each great lake there were hundreds of minor lakes, all seemingly pointing towards a former general inundation of the continental interior' (Chorley *et al.* 1964: 250). Edward Hitchcock, in his outline of the Connecticut River region, refers to the existence of former lakes and the occurrence of a diluvial debacle (Hitchcock 1819, 1824). He favoured the Noachian Deluge as the source of water, as he failed to see how more local floods could transport erratic boulders uphill:

Any one who examines the passage of the Connecticut and many of its tributaries, through several mountains embraced by this sketch, will be led, I think, to the conclusion that the waters of this river once flowed over the great valley along its banks, forming an extensive lake: and also, that when this began to subside, by the wearing away of the outer barriers, other barriers would appear and produce other lakes of inferior extent.

It is no argument, as some have thought, in favour of such a supposition, that so much rock occurs in this basin which is evidently a recomposition of the detritus of older formations; and that organic remains are found in these rocks. For every geologist knows that all this must be referred to a period anterior to that, in which the last grand diluvian catastrophe happened to the globe and left our continents in their present form. Nor is the mere occurrence of masses of stone, evidently rounded by the attrition of running water, any evidence in favour of this hypothesis; for we must look for the cause of this also, as far back at least as the Noachian deluge.—No current of water with which we are acquainted is sufficient to transport such masses of rock in the situations in which we find them: 'for though we can readily conceive how the agency of violent currents may have driven these blocks down an inclined plane, or, if the *vis a tergo* were sufficient, along a level surface, or even up a very slight and gradual acclivity, it is impossible to ascribe to them the Sisyphean labour of rolling rocky masses, sometimes of many tons in weight, up the face of abrupt and high escarpments'. Rounded masses of rock may however occur under such circumstances as to show them to have been removed by currents posterior to the deluge. (Hitchcock 1824: 16–17)

Later, in a survey of Massachusetts, Hitchcock expounded similar views, claiming that the generous covering of drift was laid down during the Flood (Hitchcock 1833). Hitchcock's catastrophist views were, for their day, fairly advanced. He held that geological processes had been more intense in the past. He believed the North American continent

to have been elevated above the ocean, not little by little, but by a few paroxysmal efforts of volcanic force. Since this force, as acting during the past four thousand years, seemed too feeble to result in the elevation of a single mountain chain, so, he argued, it must have been more energetic in previous epochs.

The fact that the older rocks are more distorted and highly meta-morphosed than the younger was thought to indicate also the greater intensity of the earlier agencies. Singularly enough, the near-shore origin of beds of conglomerate was not realized, the occasional occurrence of such among sedimentary rocks being thought due to the 'occasional occurrence of powerful debacles of water,' the like of which cannot be produced by any causes now in operation. (Merrill 1924: 146)

References to the Flood occur again and again in geological reports of the time. Elisha Mitchell, in 1827, in a paper on the low country of North Carolina, attributed the burial of the bones of elephants and mastodons to Noah's Flood. Abraham Gesner, in a volume on the geology and mineralogy of Nova Scotia published in 1836, identifies the Flood as being responsible for the formation of drift and coal (see Merrill 1924: 176). Gesner was an extreme catastrophist. This is evident in Merrill's summary of his views:

Discussing the fragments of slate and the masses of quartz rock and granite that were found scattered over the surface of the Red sandstone, and even entering into its composition at great depth, he argued that their shape demonstrated that they had been transported by the efforts of mighty currents. From this fact he conceived that similar causes had operated upon the surface of the earth at separate and distinct periods of time, one period having produced the ingredients of the newer rocks, which in their turn had been evidently denuded by the rapidity of overwhelming floods.

The giant bowlders, sometimes found on the very hilltops, he recognized as erratics, but could not believe them to be due to flood action. 'They have doubtless been thrown upwards and left cresting the highest ridges, by volcanic explosions that have taken place since the great general inundation of our planet.' The general phenomena of the drift, however, he thought to have been almost certainly the effect 'of an overwhelming deluge which at a former period produced those results now so manifest upon the earth. Not only hath the granite sent its heralds abroad, large blocks of trap are also scattered over the soil of Nova Scotia far from their original and former stations.' (Merrill 1924: 177–8)

Charles T. Jackson, head of the State Geological Survey of Maine, expresses in his annual reports decidedly diluvial views on the origin of the glacial deposits:

The 'horse-backs' (ridges of glacial gravel) were regarded by him as of diluvial material, transported by a mighty current of water which, as he supposed, rushed over the land during the last grand deluge, accounts of which had been handed down by tradition and preserved in the archives of all people. 'Although,' he wrote, 'it is commonly supposed that the deluge was intended solely for the punishment of the corrupt ante-diluvians, it is not improbable that the descendants of Noah reap many advantages from its influence, since the various soils underwent modifications and admixtures which render them better adapted for the wants of man.' 'May not the hand of benevolence be seen working even amid the waters of the deluge?' (Merrill 1924: 622)

Such a statement as this could almost have been written by John Ray. Its accuracy, however, is questionable: Merrill wonders whether the 'hard-fisted occupants of many of Maine's rocky farms would be disposed to take so cheerful a view of the matter' (1924: 622). Jackson expresses the same view in his report on the geology of Rhode Island which appeared in 1839:

There can not remain a doubt but that a violent current of water has rushed over the surface of the state since the elevation and consolidation of all the rocks and subsequent to the deposition of the Tertiary clay, and that this current came from the north. . . . Upon the surface of solid ledges, wherever they have recently been uncovered of their soil, scratches are seen running north and south, and the hard rocks are more or less polished by the currents of water which at the diluvial epoch coursed over their surfaces, carrying along pebbles and sand which effected this abrasion, leaving striae, all of which run north and south, deviating a few degrees occasionally with the changes of direction given to the currents by the obstacles in its way. (Quoted in Merrill 1924: 622–3).

Many other early American geologists wrote in a similar vein. The works of Keating, Fairholme (an English visitor to America), Rogers, and Gibson will be discussed in Chapter 9, which considers the role of debacles in landscape development. But enough has been said of this period of geological thinking in North America for the reader to appreciate that, by and large, the European system of diluvialism was transported to America, where it was adopted as the ruling theory to explain drift, erratics, and much of the topography.

7

Gradualistic Diluvialism
Floods and a non-violent history of the Earth

Precursors of gradualistic diluvialism

It is a mistaken belief held by many modern geologists, and an error found in many textbooks, that uniformitarianism commences with James Hutton's theory of the Earth which was presented as two papers and two books at the close of the seventeenth century (Hutton 1785, 1788, 1795; Playfair 1802). A number of historians of geology have recently corrected this erroneous idea. Discussions of changes of the surface of the Earth, even as early as the fifteenth century, involved a consideration of minor, everyday changes proper to the sublunar region, as well as major changes which caused the Earth's features to alter from one age to another (Kelly 1969: 219). Sublunary processes—that is, ordinary, everyday processes of nature—had been recognized by Aristotle in his *Physica* (Aristoteles 1930 edn.). They consisted of alteration, generation and corruption, growth and decay. To explain how sublunary processes could bring about major changes of the Earth's surface was no easy job. It was essentially the task Hutton and Lyell took on, but others had essayed it before them. Aristotle, for instance, in his *Meteorologica*, suggested that land and sea could change places in coastal regions owing to the drying action of the Sun (Aristoteles 1931 edn.). Valerio Faenzi, in his *De montium origine* (1561), suggested that water, running along watercourses over the Earth's surface, could slowly fashion mountains: a swift stream could, in time, change a plain into a valley with mountains on either side (see Adams 1938: 348–57). Joannes Velcurio, in a commentary on Aristotle's physics in 1588, says that mountains may be formed naturally by wind, water (floods), man, earthquakes, and giants (Eyles 1969: 221). Continual changes at the surface of the Earth were also allowed by Loys le Roy (1594). He accepted that land and sea could change places, that rivers and fountains could dry up, that the vegetation of a tract of land could change, and that mountains could be reduced to plains and plains raised to form mountains. He seemed uncertain as to the causes of the changes, listing earthquakes, heat, wind, water, air, fire, the Sun, and the other heavenly bodies as possibilities without elaborating on the mechanisms involved (Kelly 1969: 222). A more satisfactory explanation of the interchange of land and sea, the change in soil, and the building and levelling of mountains, was proffered in 1634 by Simon Stevin in his 'Second book of geography' (Kelly 1969: 222). Stevin identified wind and water as the chief agents of change in all three processes. Nathanael Carpenter went as far as to suggest that the irregularities in the Earth's surface were continually and gradually evened out by processes of erosion and deposition (Carpenter 1625). Later, Robert Hooke, whose writings were often vague and diffuse on matters geological, showed that he had a shadowy understanding of the geological cycle, and that he believed all things in nature to be in a state of flux and yet to hold a balance (Davies 1964, 1969). A balance of things in the natural world was also seen by John Ray. In his *Three physico-theological discourses* (1693), following the theory of Aristotle and Anaximenes, Ray proposed continuous changes in the position of land and sea, an overall balance being maintained so that the area of continents and oceans is always roughly the same. Ray also believed that catastrophes could occur: he thought that wide continental flats and deserts were the product of the occasional escape of subterranean waters which led to gigantic floods. None the less, the constancy of nature was a belief most dear to him; and, his vision of 'a world ceaselessly shaped and reshaped by restless waters' was adopted by Buffon as a starting-point in the development of his theory of the Earth (Fellows and Milliken 1972: 69). Buffon's work is particularly interesting, for it is one of the first attempts to explain Earth history without recourse to catastrophic events, and points the way towards uniformitarianism. In fact, Buffon's theory of the Earth does include some catastrophes, and it certainly cannot be interpreted in strict uniformitarian terms. But it is an important milestone on the road to uniformitarianism.

The most significant feature of all these early precursors of Hutton and Lyell is that the possibility was raised that the Earth had not remained completely unchanged since the Creation, but had changed—rather quickly, given the then calculated age of the Earth—owing to the agencies of wind, rain, sea, Sun, and earthquakes. But it was Lyell who so cogently argued that ordinary processes of nature, acting gradually and gently over long periods of time, were capable of slowly remodelling the Earth's topography.

The marine erosion theory

Lyell's uniformitarian model of landscape denudation

Lyell, the arch-uniformitarian, tackled a wide range of geological problems. Of special interest to the study of diluvialism is his theory attributing topography to marine erosion. Lyell subscribed to the view that the Earth's topography was fashioned by the normal action of waves and currents when the sea stood at a higher level. In proposing this idea, Lyell was reaffirming the view of Gabriel Plattes, an English writer on matters agricultural, who recognized the efficacy of the sea as an agent of erosion. In two books, *A discovery of infinite treasure, hidden since the world's beginning* and *A discovery of subterraneall treasure*, both published in London in 1639, Plattes submitted that the whole of the British landscape had been fashioned by the sea during a submergence, but that this submergence was not to be identified with the Noachian Flood. He proclaimed that the hills and dales 'doe shew plainely the worke of the water, even as the claw of a bear, or a lion, doth shew by his print that a bear or a lion hath been in such a place', and that

all England hath bin sea; by the hills, and dales, and unevennesse of the ground; being evidently graven by the water, whose propertie is to weare the ground deepest, in such places where the earth is most loose, as it is in all vallies; and to spare it most, in all rockie and firme grounds; of which sort the mountains are. (Quoted in Davies 1969: 57)

The idea of marine erosion was given an airing by Lamarck (see p. 79), but it was under the championship of Lyell that the marine erosion theory became immensely popular. In fact, Lyell's theory attributing topography to marine erosion was basically no more than a reworking of the diluvial theory of landscape development within a uniformitarian mould: 'Lyell merely took the diluvial theory, stripped it of its catastrophism, replaced the catastrophic agencies by such acceptable uniformitarian processes as wave and current action, and then adopted the refurbished theory as his own' (Davies 1969: 254). It was thus no coincidence that the marine erosion theory became fashionable in Britain during the 1830s, when the high tide of diluvialism was turning and retiring rapidly. Indeed, 'no sooner had diluvialism entered upon its decline, than the idea of a recent submergence arose phoenix-like in a fresh guise (Davies 1969: 254). And by the middle of the century, cataclysms were no longer regarded as the key to explaining the Earth's surface forms, except by a very few die-hards, who, to their fellow geologists, must have appeared unaccountably stubborn. Of course, the marine erosion theory was not the only theory of landform development during the first sixty years of the nineteenth century. Every geologist had his own recipe for landscape change, the basic ingredients of which included seismic shocks, diluvial torrents, submarine currents, swollen rivers during a pluvial period, gigantic convulsions and tsunami, debacles, and crustal subsidence; in addition, the role of fluvial and glacial processes was admitted, if generally seen as secondary. These alternatives to marine erosion are discussed elsewhere in this book.

The tenets of the new submergence theory were simple: the continents had recently been submerged; and during the submergence, waves and currents had fashioned the continental surfaces to produce the present topography. As a theory, it was similar to the old diluvialism in that both assumed a recent and widespread inundation of the continents. It differed from the old diluvialism, however, in rejecting the action of cataclysms, and instead seeking ordinary marine processes as agents of erosion. Whereas the old diluvialism was a catastrophic diluvialism, the submergence theory was what may be called a 'gradualistic diluvialism'. No doubt Lyell, not wishing to be confounded with the diluvialists, would object to this term, preferring perhaps gradualistic submergence. But it seems a good term to use for any theory which proposes a slow flooding of continents.

The submergence theory actually fitted into the uniformitarian, steady-state view of Earth history in a very fundamental way:

The stratigraphical record showed clearly that marine transgressions had occurred regularly throughout geological time, and once the uniformitarian concept of slow continental uplift was established, then it followed as a natural corollary that during each transgression and regression, marine waves and currents must have had the opportunity to work over vast areas of the continental surfaces, cutting cliffs, excavating valleys, and re-shaping the topography generally. (Davies 1969: 256)

As is usual when a new ruling theory usurps the role of its predecessor, new light seemed to be shed on a multitude of puzzling facts—a wide range of landscape features suddenly seemed easier to explain. For example, Lyell saw the topography of the Weald as a direct result of marine erosion associated with a submergence episode: he saw the scarps of the inwards-facing cuestas as cliffs cut by a retiring sea. Given that the marine erosionists had merely reframed the diluvial theory in uniformitarian terms, they could use the diluvialists' evidence of a recent submergence of the land to their own ends. The diluvial deposits, and particularly those containing shells, which lay as a blanket over much of the British landscape, were adduced by the marine erosionists, as they had been by the diluvialists, as evidence of a marine submergence. Indeed, Lyell plundered the diluvialists' storehouse of ideas quite wantonly. He dispensed with the term diluvium, which rang with far too many overtones of cataclysms, and replaced it with the term drift, which has of course stuck and is still in use. The word drift was chosen because it reflected what Lyell took as the source of the unsorted deposits—icebergs which, during the submergence, had drifted south from polar regions. Rafts of ice also offered a convenient explanation for the origin of the large erratic boulders which were so liberally dotted over the British Isles and Europe.

Lyell cunningly avoided having to pin down the date and duration of the last submergence. But during the 1830s, the marine erosionists held that the land had been submerged beneath the sea only once in modern times, and that during this recent submergence the present valleys were cut and the drift deposited by a fleet of icebergs. A problem with this interpretation arose when it was shown by the proponents of the new-fangled glacial theory that some of the drift deposits, mainly that portion now referred to as till, were deposited beneath sheets of ice and not from icebergs floating over a sea-covered continent. However, the shelly drifts and bedded drifts, today generally regarded as fluvio-glacial deposits, were still considered to be of marine origin. In fact, the marine erosionists circumvented the problem of the Ice Age by distinguishing two periods of submergence, one occurring before the Ice Age, the other within the Ice Age. They maintained that the first, or pre-glacial, submergence lasted a long time and was responsible for refashioning much of the present topography. The second, or glacial, submergence lasted too short a while to sculpture the land surface, but it did lead to the deposition of the shelly and bedded drifts.

Applications of the marine erosion theory

Lyell's marine erosion theory was given enthusiastic support in America by George E. Hayes. In his remarks upon the geology and topography of western New York, Hayes (1839) urges that the Noachian Flood be dropped as an explanation of landscape features, and proposes that the erosive action of the sea should be placed in its stead:

Why not then lay aside the fashion of attempting to explain such phenomena by invoking the assistance of the Noachian Deluge, or of tremendous inundations, sweeping over the tops of the highest mountains, produced 'by the flux and reflux of mighty deluges, caused by the sudden elevation of mountain chains in various parts of the globe?' Sound philosophy forbids these violent presumptions, particularly when the facts admit of explanations more consonant with the natural order of events.

The condition of a continent, gradually elevated from the ocean, whether by volcanic action, or by the expansive force of crystallization, or by any other cause whatever, would be such as to account for all the geological phenomena hitherto attributed to the mechanical action of water. Every portion of a continent thus reclaimed, must, in succession, have been the bed, and then the beach of an ocean. Every portion must have been subjected to the action of the waves and the tides, when lashed into fury by the raging storm: and for a period of time only limited by the greater or lesser rapidity of the elevatory process. (Hayes 1839: 88–9)

Hayes goes on to explain how, by the action of marine erosion, features such as mountain peaks, valleys, and terraces are formed on an emerging continent. Here is his explanation of valleys and terraces:

When any considerable portion [of the North American continent] had become permanently elevated above tide water, it would form a water shed, collecting the rain into rivulets, which, finding their way to the ocean, would cut out narrow channels for their beds. But the effect of these streams in the formation of valleys, by denuding and tearing up the rocky strata, would be insigificant in comparison with the action of the surge at those points where their waters were disembogued. As each portion of such channels would successively be exposed to their combined action, and must successively form the bed of an estuary at the valley's mouth, we can readily account for their excavation, to a greater or less extent, in proportion to the hardness of the rocky bed, to the violence of the waves and tides, and the duration of their action. In these estuaries, the comminuted materials would assume nearly a horizontal position, and when left dry, would resemble the alluvial plains or 'bottoms', which border most of our rivers. Should a sudden rise of a few feet take place, the water would at first recede; but by the action of the waves and tides on this alluvial mud, they would soon regain possession of that part of their former bed, bordering the stream to a greater or less extent. The centre of the valley would thereby be lowered; and this process being repeated, a series of terraces, or steps, would result, precisely similar to those in the valley of the Connecticut river, which Prof. Hitchcock attributes to the fluviatile action of existing streams. Valleys could thus be formed where streams of no great magnitude ever flowed, and where currents, except the ordinary ones of the ocean, never existed. (Hayes 1839: 89–90)

To Hayes, many other deposits in the landscape may be explained in like manner: 'The formation of sand banks and of gravel beds, the rounding and transportation of boulders, the formation and distribution of what we call diluvium, all admit the same simple explanation' (Hayes 1839: 90). And he boldly concludes that

Truth is said to be more wonderful that fiction; however this may be, it usually proves more simple than hypothesis. We ought not, therefore, to be surprised, if the phenomena which have led to the crude notion of a deluge, or a succession of deluges, have been produced by an agent no less active now than at any former time; an agent, as much more powerful in its action, as it is permanent in its duration. (Hayes 1839: 90)

Other examples of the application of the marine erosion theory are legion. Charles Darwin (1839) reinterpreted the parallel roads of Glen Roy, Inverness, previously attributed by John Macculloch (1817) to the strand lines of former lakes, as the strand lines of a former submergence. Andrew Crombie Ramsay (1846) showed that the denudation of South Wales had been accomplished by a transgressive sea, and interpreted the gently undulating uplands, like a number of gemorphologists since, as marine planation surfaces. Robert Chambers, in his book *Ancient sea margins, as memorials of changes in the relative level of land and sea* (1848), identified twenty-seven strand lines lying between sea level and 545 feet, all of which he thought dated from the submergence. The marine erosion theory was given a fillip the next year, 1849, when Thomas Stevenson, the father of Robert Louis Stevenson, measured the force of storm-waves rolling in from the Atlantic and breaking against Skerryvore Lighthouse off Tiree, in the Inner Hebrides, using a self-registering instrument. On 29 March 1845, during a heavy westerly gale, a pressure of 6083 pounds per square foot was registered, and the average pressure for the winter months of 1843 and 1844 was 2086 pounds per square foot (Stevenson 1849). This exciting evidence of the power of the sea was grist to the marine erosionists' mill. The power of the Atlantic storm-waves was also noted by Henry de la Beche, who fancied he saw evidence of the pounding action of the breakers on the western flanks of the Leinster mountains (de la Beche 1851). Almost all landforms were interpreted in terms of marine action. Solution pipes in chalk were the work of waves (Trimmer 1854). The limestone scars of Yorkshire were former sea cliffs:

The great inland cliffs, which are among the most striking phaenomena of Yorkshire, only differ from sea cliffs, because the water no longer beats against them. The Hambleton hills, the Wolds, no less than Giggleswick Scar, were cliffs against a wide sea. Kilnsey Crag was a promontory overhanging the primaeval loch, which is now the green valley of the Wharfe; and the mural precipices which gird the bases of Whernside, Ingleborough and Penyghent, formed the bold margins to similar branches of the sea, which extended up Chapeldale and Ribblesdale. (Phillips 1853: 11)

It was not just in England that the evidence of a recent submergence was found: Daniel Sharpe reported to the Geological Society of London in 1854 that, on a tour of the Alps, he had found evidence of elevated sea-levels at heights of 9000 to 9300, 7500, and 4800 feet (Sharpe 1856). Thus, the marine erosion theory seemed to explain almost every detail of the present landscape. The following extracts from the writings of J. Beete Jukes are typical of the views on the origin of topography then current:

Every valley and hollow, every slope of a hill, every cliff, every ravine, has been formed mainly by the action of the sea, when that portion of the land was at or but a little below its surface. It may have subsequently been modified by the action of the atmosphere, the frost, or the river, but its principle feature has been formed by the sea. (Jukes 1853: 152)

But when we feel ourselves entitled to take for granted that all cliffs at the foot of which the sea is now beating, have been produced by the erosive action of its waves, it only requires us to admit that the land may have stood formerly at lower levels, so as to allow the sea to flow over the lower parts of it, for us to see the probability that all inland cliffs, scars, precipices, valleys, and mountain passes, may have been produced in the same way.

The passes leading across the crests of the great mountain chains could have been produced by no other cause than by the eroding action of the tides and currents as the mountains rose through the sea; what are now 'passes' having then been 'sounds' or straits between islands. (Jukes 1862a: 101)

And according to the marine erosion theory, rivers and their tributaries could not have commenced, had not a previous system of valleys been formed by the agency of the sea:

Rivers form their own beds, but not their own valleys. Rivers are the results of their valleys, but they are their immediate results. The river could not be formed till after the valley, with all its tributary branches, had been marked out; but the valley could not even be marked out without the river, in most cases, simultaneously springing into existence, and commencing to form its channel or bed, and thus modify, and deepen, the complete valley. (Jukes 1862a: 105)

The marine erosion theory was *the* theory of landscape denudation until 1862, when Jukes, the author of the above quotations, wrote his famous paper 'On the mode of formation of some of the river-valleys in the south of Ireland', and took back all he had said about the action of the sea in shaping the land surface, proposing instead that fluvial processes were dominant in determining the Earth's topography (Jukes 1862b). This paper marked the belated advent of fluvialism, which soon became the new ruling theory of landscape development. By 1874, Lyell felt obliged to write of 'the vast spaces left vacant by the erosive power of [running] water' (Lyell 1874: 75). But the marine erosion theory did not surrender without a fight. A rearguard action was fought by Daniel Mackintosh (1815–91), who, in his *The scenery of England and Wales* (1869), still claimed that almost every facet of the topography of the English and Welsh landscape had been shaped by marine action during 'the great glacial submergence'. Feature after feature in the English and Welsh landscape is explained by Mackintosh in terms of marine erosion, but of all the landforms sculptured by the sea, he is most impressed by Brimham Rocks:

I have seen no inland rocks in Great Britain which seem to point so unequivocally to the action of the sea as the Brimham Rocks, about nine miles from Harrogate. They fringe an eminence, or upheaved island, partly spared and partly wrecked by the sea. A group of picturesque columns may be seen on the eastern shore of this ancient island, but the grand assemblage of ruins occurs on the north-western side. (Mackintosh 1869: 119)

Mackintosh then, to his satisfaction, disposes of the argument that Brimham Rocks are referable to weathering, and provides more details on the traces of marine denudation found in the vicinity:

First, a line of cliff, . . . extending along the western and north-western part of the risen island of Brimham for more than half a mile. A detached part of this coastline, behind Mrs. Weatherhead's farmhouse, shows a projecting arched rock with associated phenomena, which one familiar with sea-coast scenery could have no more hesitation in referring to wave-action than if he still beheld them whitened by the spray. Farther northwards the line of cliff in some places shows other characteristics of a modern sea-coast. Here an immense block of millstone grit has tumbled down through an undermining process—there a block seems ready to fall, but in that perilous position it would seem to have remained since the billows which failed to detach it retreated to a lower level. Along the base of the cliffs many blocks lie scattered far and near, and often occupy positions in reference to the cliffs and to each other which a power capable of transporting will alone explain. From the cliff-line passages ramify and graduate into the spaces separating the rocky pillars, which form the main attraction of this romantic spot. (Mackintosh 1869: 121)

Other features of Brimham Rocks—rock-basins, rocking-stones, and perforated rocks—are all attributed by Mackintosh to a marine agency. And in conclusion, he says

As we gaze on this wonderful group of insular wrecks, varying in form from the solemn to the grotesque, and presenting now the same general outlines with which they rose above the sea, we can scarcely resist contrasting the permanence of the 'everlasting hills' with the evanescence of man. Generation after generation of the inhabitants of the valleys within sight of the eminence on which we stand, have sunk beneath the sod, and their descendants can still behold in these rocky pillars emblems of eternity compared with their own fleeting career; but fragile, and transient, compared with the great cycle of geological events. Though the Brimham Rocks may continue invulnerable to the elements for thousands of years, their time will come, and that time will be when, through another submergence of the land, the sea shall regain ascendancy of these monuments of its ancient sway, completing the work of denudation it has left half-finished. (Mackintosh 1869: 123–4)

Mackintosh's enthusiasm is infectious, his eloquence persuasive (if a little flowery for modern tastes). He brings the landscape alive, adding zest to a subject which, if seen through the eyes of a fluvialist, is rather bland. Perhaps ramblers, standing at Brimham Rocks, would have been convinced by Mackintosh's arguments. Had Lyell stood there, he too would probably have accepted Mackintosh's interpretation. But by 1869, most physical geologists had rejected the marine erosion theory, and Mackintosh's views were disregarded.

However, Mackintosh's rearguard action on behalf of the marine erosionists was not fought in vain. Much of the geomorphological research in the first half of the twentieth century in Britain focused on denudation chronology and the changing level of the sea (Brown and Waters 1974). The denudation chronologists hunted out marine planation surfaces, chiefly in south and east Britain. Foremost among them were S. W. Wooldridge, D. L. Linton, and A. Austin Miller. These men inspired a new generation of geomorphological researchers, including E. H. Brown, who, during the 1950s, took the search for surfaces to Wales and elsewhere. Studies of pre-glacial and glacial sea-levels were apparently revitalized in 1935 by a series of lectures published under the title of *The changing sea level*, given by Henri Baulig, on a visit to London. Many British geomorphologists still give the marine erosion theory some currency: a belief in stepped marine platforms still persists, supported by strong evidence of Tertiary and Quaternary transgressions of the sea, and set against a sound and detailed understanding of the tectonic development of the British Isles (George 1974). The effects of the Calabrian marine transgression on the backslope of the Chiltern Hills have as recently as 1986 been re-evaluated using modern morphometric and sedimentological techniques (Moffatt and Catt 1986).

Marine transgressions, ice, and Earth tumble

The causes of glacial submergence and marine transgressions were not of prime concern to the marine erosionists. But during the second half of the nineteenth century, a number of theories emerged which purported to explain how the land could become deeply submerged beneath the sea. Two rather distinct groups of theories surfaced, one group explaining the swings of the sea level associated with the growth and decay of ice-sheets, the other group explaining the slow transgressions of the sea over continental lowlands. Both these groups of theories have been much refined since their inception, and have now become sophisticated indeed. It is to the flooding of the land by glacio-eustatic changes of sea level, and by marine transgressions generally, that discussion will now turn.

Floods and glacio-eustatic changes

Between about 1840 and 1860, the view held by the espousers of the glacial submergence theory was this: the glacial period was a time when valley glaciers occupied the high, mountainous regions, while the lower lands were submerged beneath a deep sea, dotted with floating icebergs, each depositing drift, dropping erratic boulders, and scratching and polishing exposed submarine rocks. During the 1860s, this glacial submergence theory gradually yielded to the land-ice theory, expounded some two decades before by Louis Agassiz (1840), which envisaged that a single ice-sheet of enormous extent had lain on much of northern Europe during the Ice Age (see Imbrie and Imbrie 1986). Two important links between the ice-sheets and sea level were established in 1865. Firstly, Searles V. Wood, in a response to a theory proposed by James Croll to account for glacial submergence (Croll 1865), pointed out that a large ice-sheet would lock up enough water to reduce the general level of the sea:

Without expressing any opinion as to the soundness of the views contemplating the existence of an ice sheet ranging up to 7,000 feet in thickness, it seems to me that a result, precisely the opposite of a glacial submergence would be the consequence; since as the sea is the source of all water, whether in the vaporous, liquid or solid form, the abstraction of so large a proportion from the fluid state, and its accumulation in a solid form over the higher latitudes, must necessarily have reduced the general sea-level, and left great areas of its shallower parts in the state of land. (Wood 1865: 297)

Thus was conceived the glacio-eustatic theory. The second link between sea level and sheets of ice was established by Thomas Jamieson. In a paper read to the Geological Society, Jamieson explained that a heavy sheet of ice would cause the land it rested on to sink, and that when the ice had melted, the land would bound back (Jamieson 1865). Thus was born the isostatic theory. These two theories caught on quickly and have proved highly productive:

The general, or almost general, acceptance of the occurrence of Ice Ages and of changes in sea level dates back to about 1870. Soon a few geologists were correlating oscillations in sea level with fluctuations in climate and a start had been made on a complex inter-relationship which still baffles geologists. (Chorley *et al.* 1964: 452)

Although the complex relationship between climate and sea-level change was indeed still a source of bafflement in 1964, the observational evidence gathered, and the development of mathematical models, since that date has led to an enormous advance in the understanding of glacio-eustatic changes. A detailed discussion of this subject is beyond the scope of the present book, but the interested reader is directed to Denton and Hughes 1983, Denton *et al.* 1986, and Huggett (in preparation).

Earth tumble and changing sea level

In 1848, a process was identified by John Lubbock which could cause the submergence of continents beneath the oceans, owing to a transgression of the sea in opposite quadrants of the globe. Lubbock suggested that the Earth's axis of rotation might change its position within the Earth. This process involves the entire globe tumbling about its fixed axis of rotation (Lubbock 1848). The possibility that the Earth might slowly tumble was discussed as early as the fifteenth century by Alessandro degli Alessandri (see p. 27), and in the seventeenth century by Robert Hooke (1688) and Isaac Newton. In his *Principia mathematica* (1687), Newton writes

But let there be added anywhere between the pole and the equator a heap of new matter like a mountain, and this by its perpetual endeavour to recede from the centre of its motion will disturb the motion of the globe, and cause its poles to wander about its superficies, describing circles about themselves and their opposite points. (Newton 1729: 1st Section, Proposition 66, Theorem 26, Corollary 22)

Newton refers to this polar motion as 'the evagation of the poles', but geographical pole shift and Earth tumble are more succinct terms for it. Lubbock, in his article communicated to the Geological Society of London by Charles Lyell, argues that

it is unlikely that when the earth was first set spinning, the axis of rotation should exactly coincide with the axis of figure, unless indeed it were all perfectly fluid. We may however go back to some time less remote, and suppose the axis of rotation not coinciding with the axis of figure, and the earth in a semi-fluid state, or rather, in consequence of the different degrees of fusibility of different substances, partly solid in irregular masses and partly fluid. We then advance to another period more recent in which the earth consisted of land and water, and was suited for the support of animal life. We may if we please consider this as the original state. The only hypothesis I wish to insist upon as essential is, that the axis of rotation had not the same geographical position as at present. (Lubbock 1848: 5)

He explains his hypothesis with the aid of a diagram (Figure 7.1a), in which the solid part of the Earth is, for simplicity, assumed to consist of a solid, spheroidal nucleus revolving about the axis CP. Under these conditions, the ocean water would be thrown into the position IKLM about the equator, the variation in depth corresponding to the polewards decrease in the angular velocity of rotation (Figure 7.1b). Were the Earth to rotate about another axis at some time, then the water would occupy a position about the new equator. Some areas, which had previously been land, would become sea, and other areas, which had previously been sea, would become land. With the assistance of another diagram (Figure 7.1c), Lubbock continues his argument, introducing climatic changes:

Now suppose a point situated at D with latitude QCD, revolving about the axis CP and submerged, were after a lapse of time to revolve about an axis CP' and having latitude DCQ', it would cease to be submerged, but at the same time would be in a colder climate, which is consistent with what you find takes place in Europe . . . ; but if we consider what takes

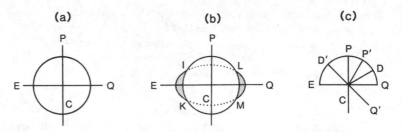

Fɪɢ. 7.1. Lubbock's diagrams demonstrating the effects of an internal change in the Earth's axis of rotation. (From Lubbock 1848)

place at the point D′ situated at a distance of 180° longitude from D, we find precisely the reverse: primitively dry the point D will become covered by sea, and will acquire a hotter climate. The countries differing in longitude from us by 180° are at present submerged by the Pacific. (Lubbock 1848: 5)

Lubbock was aware that astronomers, such as Laplace, had declared that the Earth's axis could not move in the manner he proposed. He pointed out, however, that Laplace in his analysis had not considered the dislocation of strata by cooling, or the friction of the Earth's surface, and concluded that the Earth's surface has been inhomogeneous enough to make a change in the rotation axis a possibility. A similar hypothesis to Lubbock's, reported in the briefest of notes, was offered by William Devonshire Saull at the same meeting of the Geological Society that Lubbock's paper was presented. Both Lubbock's and Saull's papers were discussed at some length, in hesitantly supportive tones, by Henry de la Beche in his anniversary address as President of the Geological Society of London (de la Beche 1849).

The debate over geographical shifts of the Earth's rotation axis warmed up again during the late 1870s. There was, for example, an exchange of views between the Reverend E. Hill, a believer in the stability of the axis, and the Reverend Osmond Fisher, an advocate of axial shift, in the *Geological Magazine* during 1878 (see Hill 1878*a*, 1878*b*; Fisher 1878*a*, 1878*b*). John Evans, in his anniversary address as President of the Geological Society of London, asked mathematicians if the elevation of a certain tract of land would not carry the Earth's axis of figure 15° or 20° away from its present position, and whether the axis of rotation would not eventually coincide with the axis of figure; in other words, he wondered whether a deformation of the Earth might not lead to a change in the position of the rotation axis within the Earth, and a change in obliquity as a final result (Evans 1876). Twisden (1878) answered Evans, calculating that the deformation envisaged by Evans would shift the axis of figure by a mere 10'. He also indicated that, if the axis of rotation and axis of figure were to separate, two vast tide-waves would sweep the Earth, submerging the equator every 150 days to a depth of six miles or more. At this point, George Darwin (1877, 1880), William Thomson (1876), and James Clerk Maxwell (1890) entered the debate, arguing with much force that anything other than a minute internal shift of the Earth's rotation axis was theoretically impossible if the Earth were solid, which they thought it was.

A way of circumventing the theorists' objections to geographical pole shift had been recognized a decade earlier by John Evans. In a paper read before the Royal Society in 1866, Evans offered an hypothesis which showed why a newly uplifted mountain mass would tend to travel towards the equator, moving the rest of the crust along with it. This may be the first time that the notion of crustal displacement was put forward. Evans's hypothesis is founded on the assumption that the globe consists of a comparatively thin crust with an internal fluid nucleus, over which the crust is free to slide when, owing to geological causes, the equilibrium is upset (Evans 1876). A similar suggestion had been made by Benjamin Franklin in two letters, one written in 1788, the other in 1790 (quoted in Merrill 1924: 13–14). Evans's thesis was strongly supported by Osmond Fisher (1878b), and was restated by the German writer, Carl Freiherr Löffelholz von Colberg, in 1886. Crustal rotation would cause the same pattern of sea-level changes as that resulting from Earth tumble.

Evidence of slow tumble: the marine transgression cycle

By the early twentieth century, the idea of slow Earth tumble was widely discussed by German geologists, despite the theorists' caveats. Reibisch (1901), Damian Kreichgauer (1902), and Alfred Lothar Wegener (1915) all suggested that evidence for past geographical shifts of the poles should be found in the marine transgression cycle. They reasoned, as Lubbock had done half a century before, that the shape of the Earth is ellipsoidal. If the poles shift internally, in other words if the Earth tumbles, then it will take a long time for the solid Earth to adjust to the new position of the axis. On the other hand, the oceans, being fluid, will adjust at once. The result is a global change of sea level, with transgressions and regressions in different quadrants of the Earth. Wegener used a sketch (Figure 7.2) to explain this phenomenon:

Since the ocean follows immediately any re-orientation of the equatorial bulge, but the earth does not, then in the quadrant in front of the wandering pole increasing regression or formation of dry land prevails; in the quadrant behind, increasing transgression or inundation. Since the equatorial radius of the earth is about 21,000 m greater than the polar, then with the 60° polar wandering between the Carboniferous and the Quaternary, if it was accompanied by an equal amount of internal axial shift, Spitsbergen would have had to rise about 20 km above the surface of the sea, and central Africa would have had to sink a similar amount below, if the earth had retained its shape. Naturally, the latter cannot have been the case, because the possibility of large axial shifts depends on its re-orientation flow. However, the adjustment probably involved a lag of the order of magnitude of 100 m behind the immediate re-setting of the ocean level, and this must have shown up as transgression cycles. (Wegener 1929: 159)

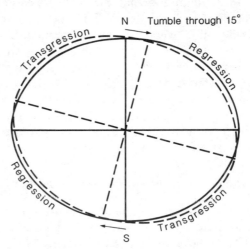

Fig. 7.2. Wegener's sketch of transgressions and regressions of the oceans owing to true polar wander. In the example depicted, the Earth undergoes a slow tumble through 15°. (After Wegener 1929)

Wegener argued that in the quadrant ahead of the migrating pole evidence of increasing regression should be found in the strati-graphic record; the reverse should apply in the quadrant in the rear. He believed that the stratigraphic sequence between the Devonian and Permian contains such evidence. A more recent review of these ideas is found in Fairbridge (1961).

Theoretical and observational breakthroughs

Wegener's ideas on the marine transgression cycle, though interesting, were not supported by a firm, modern theory of Earth tumble. A theory in which large geographical shifts of the poles were allowed was promulgated in 1955 by the British astronomer Thomas Gold. From an analysis of minor rotational perturbations other than precession, Gold demonstrated that plastic flow of the Earth must be possible, and that even a slight redistribution of the Earth's mass will cause the poles to shift, despite the stabilizing effect of the equatorial bulge. For instance, if South America were raised by 30 m over a few million years, the poles would rotate by 90° at the rate of about 1° per millenium. Gold thought that generally this does not happen, but was tempted to suggest that there have been a few occasions when the Earth's axis has been 'free' and swung around.

Since Gold's paper, and with advent of palaeomagnetic data, the hypothesis of true polar wandering has been taken seriously and there is now a consensus that it does occur (see Goldreich and Toomre 1969; Hargreaves and Duncan 1973; Jurdy and Van der Voo 1975; Jurdy 1981, 1983; Morgan 1983; Andrews 1985). Indeed, the debate is not now about whether true polar wander occurs, but about what exactly it is that moves when it does occur. The three currently testable possibilities are that the lithosphere moves, that the mantle moves, and that the lithosphere and mantle move together. Goldreich and Toomre (1969), in their theoretical analysis of polar wandering, lend support to the hypothesis that large angular displacements of the Earth's rotation axis relative to the entire mantle have occurred on a geological timescale owing to the gradual redistribution (or decay or manufacture) of density inhomogeneities in the Earth. The view that the mantle moves by itself, by the process of mantle roll, was proposed by Hargreaves and Duncan (1973) and is supported by the work of Morgan (1981,

1983). The view that the whole Earth, or the lithosphere and mantle together, tumble is favoured by Andrews (1985) as the most likely cause of true polar wander.

Studies of true polar wander are not purely theoretical. Using the palaeomagnetic data available for Phanerozoic rocks, Andrews (1985) has traced the true wander of the poles during the Cenozoic and Mesozoic by looking at the relative motion between the framework of hotspots, assumed to be fixed in the mantle, and the geomagnetic poles as deduced from worldwide palaeomagnetic data. She found that the poles have wandered 22° (± 10°) over the last 180 million years, giving a rate of true polar wander of 1° per million years. She also found episodes of slow wander, less than 2 cm/year, in the periods 5 to 50 and 115 to 160 million years ago; and episodes of rapid wander, in the range 8 to 10 cm/year, in the periods 50 to 65, 85 to 115, and 160 to 180 million years ago. She concludes that, as there is an observed relative motion between the rotated palaeopoles, or palaeorotation axis, and the hotspot reference frame, the whole Earth, or at least the mantle and the overlying lithospherical plates, tumbles: the geographical poles defined by the rotation axis change over time. The suggested cause of this process is the redistribution of mass in the plastic mantle from any region of high density towards the equator. Independent work by Courtillot and Besse (1987) analysed the shift of the entire mantle relative to the Earth's rotation axis over the last 200 million years to yield the pattern of true polar wander. Fast polar wander, averaging between 4 and 5 cm/yr, occurred from 200 to 170 million years ago. There was then a stillstand lasting 60 million years. A fast phase ensued from 110 to around 40 to 30 million years ago, when a sharper cusp (almost a hairpin) occurred with no indication of retardation (Figure 7.3). These results are similar to Andrews's findings, but the polar wander curve is less erratic.

Whatever the cause of Earth tumble, the evidence from polar wander studies strongly suggests that it has taken place. And if it has occurred then it will have led to marine transgressions of the nature envisaged by Wegener. Even a small shift in the position of the geographical poles will lead to the flooding of large areas of continental lowland. To elucidate this point, it is instructive to compute the change in sea level at a particular point on the Earth resulting from polar displacement of a given degree. The sea-level change is readily calculated from the equation for an ellipsoid

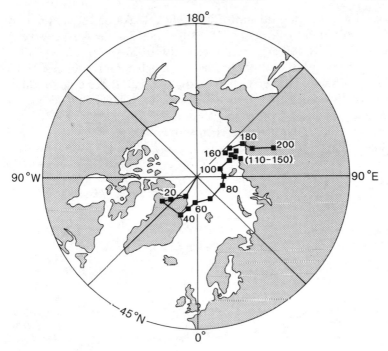

FIG. 7.3. True polar wander as computed by Courtillot and Besse. The trace represents global motion of the mantle with respect to the Earth's rotation axis. (After Courtillot and Besse 1987)

(Weyer 1978):

$$x^2/a^2 + y^2/b^2 = 1.$$

The difference in sea level, d, in metres is

$$d = \{\cos(lat_2) - \cos(lat_1)\} \times 10\,675.$$

A positive difference means that sea level is higher at the new latitude, lat_2, than at the old latitude, lat_1, and vice versa. The new latitude depends on the amount of pole shift, c, and the difference in longitude, a, between a particular point on the Earth and the meridian of pole shift. It may be computed as

$$\sin(lat_2) = \sin(lat_1).\cos c + \cos(lat_1).\sin c.\cos a.$$

Gradualistic diluvialism

A variety of cases are shown in Table 7.1. Along the meridian of pole shift, a displacement of 1° leads to a 373 m drop in sea level at latitude 45 °N and a corresponding rise of 373 m at latitude 45 °S. Bigger pole shifts naturally lead to much larger changes of sea level. A shift of 45° gives a drop of 10 675 m at the equator and a rise of 10 675 m at the poles. The biggest sea-level changes are, of course, associated with a pole shift of 90° which leads to a 21 350 m drop at the equator and a corresponding rise of 21 350 m at the poles. The changes of sea level become smaller away from the

TABLE 7.1. *Sea-level change resulting from a 1°, 45°, and 90° shift of the poles*

Old latitude in all cases (°N or S)	1° Polar shift		45° Polar shift		90° Polar shift	
	New latitude (°N or S)	Sea-level change[a,b] (m)	New latitude (°N or S)	Sea-level change[a,b] (m)	New latitude (°N or S)	Sea-level change[a,b] (m)
90	89	7	45	10 675	0	21 350
85	86	−58	50	8 659	5	21 026
80	81	−121	55	6 380	10	20 062
75	76	−181	60	3 907	15	18 490
70	71	−234	65	1 316	20	16 355
65	66	−281	70	−1 316	25	13 724
60	61	−319	75	−3 907	30	10 675
55	56	−348	80	−6 380	35	7 302
50	51	−366	85	−8 659	40	3 707
45	46	−373	90	−10 675	45	0
40	41	−368	85	−12 367	50	−3 707
35	36	−352	80	−13 682	55	−7 302
30	31	−326	75	−14 582	60	−10 675
25	26	−290	70	−15 039	65	−13 724
20	21	−244	65	−15 039	70	−16 355
15	16	−192	60	−14 582	75	−18 490
10	11	−134	55	−13 682	80	−20 062
5	6	−71	50	−12 367	85	−21 026
0	1	−7	45	10 675	90	−21 350

[a] The sea-level changes listed here occur along the meridian of pole shift. Sea-level changes at other longitudes are smaller.

[b] The table shows sea-level changes for one quadrant. Sea level in latitudinally adjacent quadrants changes by the same absolute amount but the signs are reversed; and in some cases, for instance with a 90° polar shift, all latitudes move into the opposite hemisphere.

meridian of pole shift. The values computed using the above formula are maximum values: the actual sea-level change will be less if the solid Earth yields ellipsoidally under the same forces as the ocean (Weyer 1978). Clearly, even shifts of the poles as small as one degree would lead to a gradual inundation of large areas of continental lowland producing new shorelines and burying river courses and their associated terraces. Indeed, the French geographer, Jacques Blanchard (1942), has suggested that the poles may have shifted on a cyclical basis owing to a more pronounced wobble of the Earth about its axis in the past. He showed that there were at least twelve major climatic changes in the valley of the Somme during the last ice age which are associated with changes of sea level, fossil assemblages, and human cultures. He argued that only displacements of the poles can explain this sequence of events.

Marine transgressions and geotectonics

Earth tumble is a sufficient, but not necessary, cause of marine transgressions. Another cause is a change in the volume of the mid-ocean ridge system. This cause brings into purview the question of plate tectonics, and the modern history of ideas concerning global changes of sea level (see Fairbridge 1961 and Mörner 1987 for reviews). Work on relative changes of sea level on a worldwide scale will now be considered.

Cycles of sea-level change

Relative changes of sea level on a global scale were considered by Robert Chambers in 1848, but were first explored fully by the great Austrian geologist, Eduard Suess (1831–1914). In his *magnum opus, Das Anlitz der Erde* (1885–1909, 1904–24), Suess claimed the existence of global transgressions and regressions. Perhaps the best example of such an episode of advance and retreat of the oceans over all continents, is the Late Cretaceous transgression which submerged more than half of Europe, North America, Arabia, Iran, and North India under a shallow sea (Figure 7.4). Suess's work was developed by Schuchert, Stille, Grabau, Haug, Kuenen, and Umbgrove. The researches of these pioneering

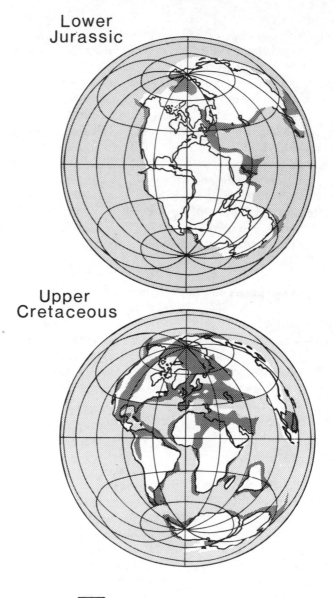

Lower
Jurassic

Upper
Cretaceous

Flooded continent

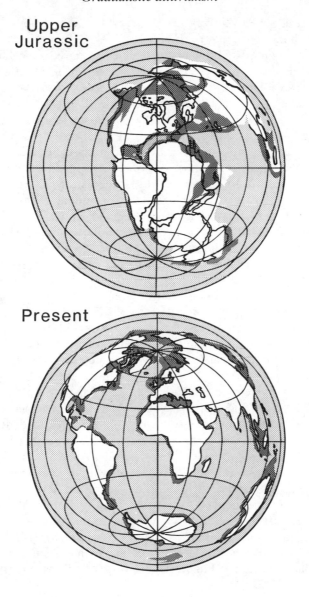

FIG. 7.4. Continental flooding during the breakup of Pangaea. (After Nance *et al.* 1988, who based the maps on the work of A. G. Smith and J. C. Briden)

geologists paved the way for the recent studies of episodic and cyclic changes of sea level (Williams 1981: 217).

From about 1950, a flood of studies appeared revealing long-term cycles of sea-level change, each cycle involving a transgressional and regressional phase. Strakhov (1949) distinguished twelve such grand cycles during the Phanerozoic with periods of 20 to 35 million years. He found that long phases of intense regression, lasting 10 to 20 million years, occurred at the start of the Cambrian and Devonian, and in the middle of the Jurassic periods, and were associated with strong phases of orogeny. Umbgrove (1947: 92), too, had associated periods of global regression with episodes of orogenesis, but many recent authors, including Johnson (1971), hold that orogenic phases lead to global transgression. Egyed (1956*a*, 1956*b*) used palaeogeographical maps of the world to calculate the area of the continents covered by water since the start of the Cambrian period through to the present. By plotting the area of continental flooding against time, he showed an apparent emergence of the continents during the Phanerozoic, on which was superimposed long-term oscillations of sea level. He distinguished 8.5 swings from transgression to regression during the past 400 million years, and suggested that their mean period of 47 million years was a reflection of the periodic accumulation and release of global tectonic forces. Doubt was cast by Wise (1974) on the method by which Egyed had calculated his sea-level curve. To check Egyed's findings, Wise used Schuchert's palaeographical atlas of North America and computed an independent curve. Wise's results showed a periodic rise and fall of sea level, with an overall period of roughly 35 to 45 million years, but without a secular decrease in freeboard (the relative elevation of the continents with respect to the oceans) through time. However, a new twist has been given to this particular line of investigation by Hallam's curves of relative sea level based on even more accurate palaeogeographical and facies maps of Russia and North America, which seem to confirm Egyed's notion of a secular decrease in freeboard during the Phanerozoic with oscillations superimposed (Hallam 1977).

Sea level and mid-ocean ridges

During the 1970s, the study of global relative sea-level changes took two separate, but connected, directions. On the one hand seismic stratigraphy was used by P. R. Vail and his colleagues at the Exxon exploration laboratories in Houston, Texas, to determine relative changes of sea level (Vail *et al.* 1977). On the other hand, the idea was explored that the worldwide mid-ocean ridge system might, by changing its volume, lead to changes of sea level.

The plate tectonic model of relative sea-level changes was first proposed by Hallam (1963), and developed by Menard (1964, 1969), Russel (1968), and Valentine and Moores (1972). It is a simple but most effective idea. In brief, it suggests that an increase in the volume of the mid-ocean ridge system would cause the oceans to overflow their basins and spill on to continental lowlands as a transgression; conversely, a decrease in the volume of the mid-ocean ridge system would cause the oceans to retire to their basins, leaving the previously flooded portions of continents dry, once the regression was complete. Hallam (1971) elaborated the idea by suggesting that if the rate of sea-floor spreading increased or decreased, then the mid-ocean ridges would become narrower or wider respectively, so causing global transgressions or regressions. This idea is supported by the finding that the rate of sea-floor spreading was uncommonly great, twice the normal rate, throughout the world in the Middle Cretaceous, with a peak at about 85 million years ago (Larson and Pitman 1972). If this increase in rate increased the volume of the mid-ocean ridge system appreciably, then a marine transgression would be expected to occur. A major transgression did occur in the Late Cretaceous, and Hays and Pitman (1973) suggested that it might have been caused by the elevated spreading-rate of oceanic plates. Other processes, including the release of juvenile water at active ridge edges, volume changes of the ocean basins owing to the differentiation of the lithosphere, variations in sedimentation, and crustal shortening, may lead to smaller relative change of sea level (Hays and Pitman 1973; Pitman 1978).

Hays and Pitman (1973) have refined the plate tectonic model of relative sea-level change so that a change in the volume of the mid-ocean ridge system can be converted into an actual sea-level change. To make the conversion from ocean basin capacity to change of sea level requires two corrections to be made (Hays and Pitman 1973). The first is a correction for the isostatic adjustment of the ocean basins relative to the continents. When the depth of ocean water increases by an amount, h, then the ocean floor subsides a distance, d. The relationship between h and d, with the density of the upper mantle given as 3.4 g/cm^3, is

$h = 3.4\,d$.

Thus the change in continental freeboard is

$h - d = 0.7\,h$.

The second correction allows for the increase in the area of the land surface covered by the ocean as the sea level rises. Roughly one sixth, or 8.5×10^7 km^2, of the Earth's surface lies between 0 and 500 metres. On the assumption that as the sea level rises, the additional area of land flooded increases linearly at the rate of 1.7×10^5 km^2 per metre rise of sea level, then the actual sea-level change, $0.7\,h$, can be calculated in metres as

$$\Delta V = hA_0 + 170\,(0.7\,h)^2/2$$

Where ΔV (km^3) is the change in the volume of the mid-ocean ridge system, and A_0 (360×10^6 km^2) is the present-day area of the oceans. Using this computational procedure, Pitman (1978) predicted the curve of relative sea-level change from the upper Cretaceous to the mid-Miocene, at which time glacio-eustatic changes became dominant (Figure 7.5). Notice in Figure 7.5 that sea level fell steadily at a rate of about 0.7 cm per year during the entire period, though phases of relatively slow and fast fall are evident. Also notice that sea level was about 350 metres higher in Late Cretaceous time than it is now. Pitman (1978) refined his model to include the effects of the deposition, and subsequent subsidence, of sediments along continental margins within several hundred metres of sea level. He was able to show that an increase in the rate of fall of sea level would require the shoreline to migrate seawards, so producing a regression; and that a decrease in the rate of fall of sea level would require the shoreline to migrate landwards, so producing a transgression. Thus, transgressive and regressive movements of the sea cannot be read directly in the curves of relative sea-level change: transgressions and regressions do not necessarily correspond to the high-stands and low-stands of sea level. Rather, they result from changes in the rate of rise and fall of sea level: a transgression occurs, either when sea level increases its rate of rise, or when it decreases its rate of fall; a regression occurs, either when sea level decreases its rate of rise, or when it increases its rate of fall. The transgression and regression which occurred during the Oligocene is thus the result of changes in the rate of rise and fall of sea level (Figure 7.5).

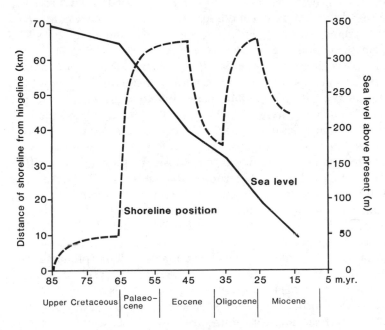

FIG. 7.5. The change of sea level (continuous line), and distance of the shoreline from the hingeline (broken line), from 85 to 15 million years ago. The hingeline is a fixed line about which subsidence of the basement rocks is assumed to occur owing to the accumulation of sediments. Notice that the Eocene transgression and the Oligocene regression are the results of changes in the rate of the fall of sea level. (After Pitman 1978)

A different method of establishing relative sea-level change is provided by seismic stratigraphy, the details of which technique are summarized by Hallam (1978). The results of modal averages of correlated regional sea-level cycles reveal a hierarchy of global cycles of relative sea-level change during the Phanerozoic. Three orders of cycles are superimposed on the sea-level curve. The first-order cycles have a duration of 200 to 300 million years; the second-order cycles have a duration of 10 to 80 million years; and the third-order cycles have a duration of 1 to 10 million years. The second-order and third-order cycles are characterized by the following sequence of sea-level change: a gradual rise, a standstill, and a rapid fall. Williams notes that the relative sea-level change curves of Wise, Hallam, and Vail *et al.*, although they were derived using different methods, have much in common: firstly, they show overall high sea levels during the early to middle Palaeozoic and the middle to late Mesozoic, suggesting a 300-million-year cycle; secondly, they exhibit more rapid, second-order oscillations of sea level superimposed on the long cycle, with durations of 10 to 80 million years and mean periods between 35 and 55 million years; and, thirdly, they all demonstrate a tendency for the boundaries of geological periods to correspond to relatively low sea levels, 'a circumstance that may be fundamental to our understanding of revolutions in the history of life' (Williams 1981: 218). As to the possible cause of long-term sea-level change, Williams points to a combination of crustal thickening by orogeny and variation in the volume, cumulative length, and spreading rates of the mid-ocean ridge system. Vail *et al.* (1977: 94) agree that geotectonic processes may be the dominant causes of long-term sea-level change, suggesting that sea-floor spreading rates and orogeny may determine the first-order cycle of sea-level change, and orogenic movements and volcanism may determine the second-order high-stands of sea level. All these long-term changes of sea level are thus tectono-eustatic in origin, and not glacio-eustatic.

Sea level and the supercontinent cycle

Another possible geotectonic explanation of relative sea-level change throughout Earth history is offered by the theory of the 'supercontinent cycle' proposed by Worsley *et al.* (1984) and Nance *et al.* (1988). According to this theory, the continents repeatedly coalesce to form a supercontinent, and then break into smaller continents, owing to the pattern of heat conduction and loss through the crust. The process envisaged is this:

if a stationary supercontinent covers some part of the earth's surface, heat from the mantle should accumulate under the supercontinent, causing it to dome upward and eventually break apart. As fragments of the supercontinent disperse, heat can be transferred through the new ocean basins created between them. After a certain amount of heat has escaped, the continental fragments may be driven back together.

In other words, we think the surface of the earth is like a coffee percolator. As in a coffee percolator, the input of heat is essentially continuous. Because of poor conduction through the continents, however, the heat is released in relatively sudden bursts. (Nance *et al.* 1988: 44)

One of the many effects of the proposed supercontinent cycle would be to change sea level in a systematic manner, owing to thermal upheaving of continents and the creation and destruction of ocean basins. The combined results of the geotectonic processes which occur during a supercontinent cycle would lead to the following sequence of sea-level change:

during the existence of a supercontinent sea level should be relatively low. As the supercontinent breaks up, sea level should rise, both because the continental fragments will stretch and subside thermally and because the breakup will replace old, Pacific-type ocean with young, Atlantic-type ocean. Sea level should continue to rise for about 80 million years, as the younger oceans make up a greater fraction of the world ocean. Then, as the Atlantic-type oceans age and expand, sea level should decline for another 80 million years or so, until the Atlantic-type oceans begin to be subducted.

When the continents begin to come together, sea level should rise, as older Atlantic-type crust is subducted. That rise in sea level should continue for another 80 million years, until the supercontinent begins to be reassembled. Then, as continents collide and the growing supercontinent is uplifted thermally, sea level should decline for about 80 million years. Once the supercontinent has been formed, sea level should remain static for another 120 million years, until the supercontinent breaks up again. (Nance *et al.* 1988: 48)

Marine transgressions and the landscape

Whatever their cause, transgressions and regressions occur. They inundate large areas of continents and, in doing so, may cause a change of climate. A transgressive sea cannot perform the catastrophic erosion that rapid floods can, but it slowly works over continents and can be expected to leave traces in the landscape, even though the traces might subsequently be buried beneath sediments. It may even be capable of producing relatively flat areas of land, which are today expressed as planation surfaces. The theory of marine planation is unfashionable. Its more recent advocates have been accused of letting their eyes deceive them. Most purely morphological evidence, we are told, is 'so ambiguous that theory feeds readily on preconception' (Chorley 1965: 151). But it is possible that transgressive seas did actually produce some of the 'surfaces' and the other features which Wooldridge, Linton, and their students saw almost everywhere in the British landscape. When Lyell proposed his marine erosion theory, he was not, it would seem, doing geomorphology quite the disservice that some modern writers have suggested.

8

Diluvialism, Debacles, and Climatic Change
Floods from lake bursts and a pluvial period

Floods and a change of climate

Swollen rivers in a pluvial period

During the 1790s, Richard Kirwan toyed with the idea of explaining modern topography in terms of fluvial erosion performed during a former pluvial period. John Carr (1809a, 1809b) and John Macculloch (1817) both developed the idea a little. George Julius Poulett Scrope mentioned in his *Considerations on volcanos* (1825) that many fluvial features in the landscape were shaped during a different rainfall regime. But the notion of greatly increased fluvial activity during a past pluvial period really came to the fore in 1859. In that year, Joseph Prestwich visited a site in the Somme valley near Abbeville which had been excavated in 1838 by Jacques Boucher de Perthes. The excavations had revealed forty feet of gravel in which artefacts were associated with extinct animal remains (Daniel 1959, 1981). Having investigated the site, Prestwich concluded that the gravel deposits were locally derived from within the Somme valley (1862–3a, 1862–3b, 1863–4, 1864). He noticed that there were two gravel suites at different heights. The upper gravel terraces were older and coarser, but none the less produced, claimed Prestwich, by ordinary fluvial processes. However, he thought that they could not have been produced by the fluvial regime which is found in the Somme valley at present. Rather, he said, they were formed under the severer climate of the Pleistocene, when seasonal melt of local ice-sheets would have produced powerful floods, particularly in rivers fed by mountain streams:

These conditions, taken as a whole, are compatible only with the action of rivers flowing in the direction of the present rivers, and in operation before the existing valleys were excavated through the higher plains, of power and volume far greater than the present rivers, and dependent upon climatal causes distinct from those now prevailing in these latitudes.

The size, power, and width of the old rivers is clearly evinced by the breadth of their channels, and the coarseness and mass of their shingle beds; whilst the volume and power of the periodical inundations are proved by the great height to which the flood-silt has been carried above the ordinary old river-levels—floods which swept down the land and marsh shells, together with the remains of animals of the adjacent shores, and entombed them in the coarser shingle of the main channel, or else in the finer sediment deposited by the subsiding waters in the more sheltered positions. As the main channel was deepened from year to year by the scouring action of the rivers, the older shingle banks were after a time left dry, except during floods, when they became covered up with the flood-silt, which, extending also over the adjacent land and shores, was there deposited directly upon the rocky substratum. As the channel became deeper, and the tributary valleys partook of the same erosion, they, being out of the main river-current, tended especially to receive thick deposits of flood-silt (Loess), while the higher grounds were left permanently dry. (Prestwich 1864: 286–7)

An alternative explanation of the origin of the gravels, still involving a change of climate and a concomitant change in the fluvial regime, was proffered by Alfred Tylor. Instead of regarding the gravels as two separate suites, he suggested that they were a single formation of roughly the same age; that the valley was cut before the gravels were laid down; and that, just before historical times, heavy rainfall associated with a 'pluvial period' caused the rivers to rise to the level of the highest gravels, which were then deposited (Tylor 1866, 1868). Prestwich was not averse to the notion of a pluvial period, but he stood by his claim that the terrace gravels were deposited at two different times. George Greenwood was less favourably disposed towards Tylor's suggestions, and felt moved to pen the following invective:

Mr. Tylor, while he considerately spares us a 'gravel period', creates a bran new period of his own—a pluvial period. With this implement (notwithstanding that 'a valley of the Somme had assumed its present form prior to the deposition of any of the gravel or "loess" now to be seen there'), he floods the valley 'eighty feet above the present level of the Somme'. These prodigious bodies of water do not in the least erode the soft chalk sides, or the bed of the valley, but, on the contrary, they deposit the gravel terraces at their high-water mark. Flints, therefore, in the pluvial period, must have been lighter than water, and must have floated on the surface to their present position. (Greenwood 1877: 166–7)

Fresh research carried out by Tylor, this time on deltaic deposits, led him to reaffirm his concept of a pluvial period (Tylor 1869). He found that the sediments at the bottom of a delta are coarser than those at the top. The coarser, bottom sediments, he concluded, were deposited by fast-flowing, swollen rivers during a pluvial period, when the sea level was about six hundred feet lower than at present. The vestiges of a past pluvial regime found at the base of deltas corresponded to the gravels laid down on valley-sides by rivers swollen to about eighty feet above their normal level. In a later article, he bolstered his thesis with recent findings on sea-level fluctuations, showing how such fluctuations were reflected in the structure of the Ganges, Mississippi, Po, and Volga deltas (Tylor 1872). He also restated his position on the Somme valleys being products of a pluvial period: 'No one can see the great valley of the Somme, or the Dover Valley, without being convinced that in the Quaternary Period these wide and deep valleys, excavated out of solid chalk, were filled by large rivers' (Tylor 1872: 395).

The hypothesis of the pluvial period still had much currency in the twentieth century. Clement Reid, in his paper on the 'Ancient rivers of Bournemouth' (1915), accounts for the spreads of gravel in the region in terms of climatic change:

With an Arctic climate conditions would be entirely altered [from present conditions], even if the combined rainfall and snowfall were no greater than now. During the long winter, erosion would almost cease, but during the spring the rainfall and the melting of the snow accumulated during several months would cause floods such as we now never see in the south of England. Not only so, but the formation of bottom or anchor-ice, and the floating off of large cakes of ice laden with stones would clear away and sweep down the streams large boulders such as no flood nowadays can move. (Reid 1915: 78–9)

Similarly, Henry Bury, talking of the spreads of gravel on the Bournemouth plateau, explains that

No one indeed supposes that such small rivers as we see at the present day could have produced these huge sheets of gravel; those of former days must have been, at certain seasons at least, much larger and more violent; and the most probable cause for such increase in size is to be found in the melting of snows at the close of one of the several Glacial Periods, which we know occurred in Pleistocene times. There is no evidence of actual glaciers south of the Thames, but there was probably a great accumulation of snow, and a deeply frozen soil; and when glacial conditions began to pass away, there must have been torrential floods every summer, and perhaps sheets of half-frozen sludge creeping down the hill-sides. (Bury 1923–4: 76)

All this relatively early work on terrace gravels and pluvial periods was, in Britain at least, rather neglected during the years when the 'interglacial sea-level' hypothesis was the ruling theory (Jones 1981: 142). However, emphasis has over the last twenty years shifted back to the relationship between gravel deposition and changes in palaeodischarge, as the key to understanding terrace formation. Also, climatic change and its effect on river discharge during the Pleistocene has been shown by Dury to be a likely cause of meandering valleys (Dury 1953, 1964, 1965, 1969). Dury noticed that many meandering streams are at present manifestly underfit: it is self-evident from looking at them that their bends are smaller than the bends of the valleys that they occupy. He found that the wavelength of valley bends is between about five and ten times the wavelength of present-day stream meanders. He suggested that the valley bends were cut when the channel-forming discharge was far greater than it is now. By using known empirical relationships between meander wavelength and bankfull discharge, and allowing for additional hydraulic characteristics of the stream, he was able to show that a five-fold increase of wavelength requires about a twenty-fold increase in bankfull discharge. He regarded climatic change as the root cause of this increased discharge, but recognized that other factors could produce the same result:

Now although in special conditions underfitness may be due to river capture, the cessation of meltwater discharge from glaciers, or the cessation of overspill from ice-dammed lakes, most streams which are now underfit have had their channel-forming discharges reduced by climatic change. It is highly likely that increase, reduction, and renewed increase of discharge occurred repeatedly during the Pleistocene, but little is known of any but the last main episode of shrinkage. This, possibly itself interrupted by an increase, took place between about 12,000 and 9,000 years ago. This was the time of the last major transition from high-glacial to interglacial conditions. When allowance is made for the reduced air temperatures of the time it can be shown that the swollen channel-forming discharge required to shape valley meanders could be produced by an increase in mean annual precipitation to 1½ or 2 times its present value. (Dury 1969: 428–9)

Seasonal floods during the Ice Age

A number of writers have suggested that floods, possibly of cataclysmic proportions, may be associated with ice ages. In 1842, a French mathematician, Joseph Alphonse Adhémar, published a book called *Révolutions de la mer* in which he suggested that ice ages may be produced by changes in the orbital motions of the Earth. In proposing this carefully argued hypothesis, Adhémar also made the outrageous suggestion that the gravitational pull of the Antarctic ice-sheet was big enough to drain the water from the northern hemisphere and create a sea-level bulge in the southern hemisphere. He also predicted that when eventually temperatures in the southern hemisphere began to rise, the Antarctic ice-cap would melt, eaten away at its base to leave a gigantic mushroom-like structure which would eventually collapse into the ocean creating a huge iceberg-laden tidal wave that would sweep northwards and engulf the land.

Adhémar was mainly interested in the precessional cycle of the Earth; other writers have suggested that the tilt cycle may produce ice ages and lead to seasonal floods in some latitudes. An hypothesis which predicts such seasonal flooding during the last Glacial Epoch owing to an increased tilt of the Earth was put forward by an English Army Officer, Major-General Alfred Wilks Drayson. Drayson's hypothesis was communicated to the Geological Society of London by Alfred Tylor (Drayson 1871), but its first full exposition is in *The cause, date, and duration of the last glacial epoch of geology* (1873). The nub of Drayson's hypothesis centres on the fact that two well-known and generally accepted statements are, in fact, contradictory: that the pole of the heavens describes a circle round the pole of the ecliptic as a centre, keeping always at a distance of 23° 28' from that centre; and that, from historical ages down to the present, the obliquity of the ecliptic has decreased at a rate of about 45″ per century (see Huggett in preparation). Drayson points out that, if the pole of the ecliptic be the centre of the circle described by the pole of the heavens, then no variation in the obliquity could occur, and the recorded decrease in obliquity would be impossible. But, because the decrease in the obliquity seems an incontrovertible fact, there is no alternative but to discard the statement that the pole of the heavens describes a circle centred on the pole of the ecliptic. Having made that point, and having driven it home, Drayson sets about retracing the actual path of the Earth's axis using the values of obliquity recorded between AD 1437 and 1870. Given that the pole of the heavens moves at the rate of about 1° in 180 years, the data give the coordinates of the actual curve traced by the pole. Drayson calculates the 'true' path traced by the Earth's rotation axis in the heavens, and finds its centre to be located 6° from the pole of the ecliptic, at point C in Figure 8.1. He then points out that, in moving round the pole of the ecliptic, the Earth's axis will gradually change its obliquity. The curve also implies that the Earth's obliquity changes through a cycle corresponding to one passage of the rotation axis along the circle. Drayson (1873: 141) estimates that this cycle of the direction of the earth's rotation axis has a period of about 31 840 years (31 756 according to De Horsey 1911; 31 682 according to Marriott 1914), and he charts the climate of the Earth as it changes owing to his proposed 31 840-year cycle of obliquity. He starts his account at position M (Figure 8.1), which would have occurred in 13 700 BC (13 544 BC according to Marriott 1914) when the obliquity was 35° 25′ 47″ (Figure 8.2). At this time, the Arctic Circle would have extended to a latitude of

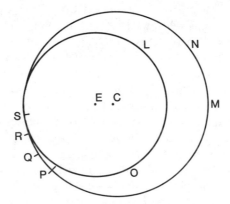

Fɪɢ 8.1. Drayson's diagram showing the 'true' course of the Earth's rotation axis as traced in the heavens. The point E is the pole of the ecliptic. (From Drayson 1873: 128)

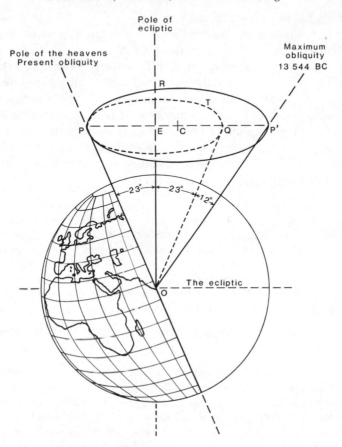

Fig. 8.2. Marriott's diagram showing the slow conical motion of the Earth's axis which causes the precession of the equinoxes, as proposed by Drayson. The dotted circle (in perspective) shows the path of the pole of the heavens, according to the usual astronomical theory. The revolution is supposed to take 25 868 years to complete. The outer circle PRP' shows the path of the pole as worked out by Drayson, and is completed in 31 682 years (31 756 years according to De Horsey 1911). Obliquity of the ecliptic is at its maximum 35° 25′ 47″, and at its minimum 23° 25′ 47″. The present obliquity (1922) is 23° 26′ 58″. (After Marriott 1914 and Barley 1922)

nearly 54°, the altitude of the Sun being 12° less than at present. North of that latitude, winter conditions would have been similar to those experienced today north of the Arctic Circle. The greater part of England and Wales would be 'covered with a mantle of snow and ice, and those creatures which could neither migrate nor endure the cold of an arctic climate would be exterminated' (Drayson 1873: 147). At the summer solstice, the Sun's altitude at 54 °N would be roughly the same as the present altitude attained in southern Spain and southern Italy. However, at a latitude of 54° the Sun would remain the whole 24 hours above the horizon at the period of the summer solstice, bringing

extreme heat to those very regions which, six months previously, had been subjected to an arctic cold. Not only would this greatly increased heat prevail in the latitude of 54°, but the sun's altitude would be 12° greater at midday in midsummer, and also 12° greater at midnight in high northern latitudes, than it ever attains now; consequently the heat would be far greater than at present, and high northern regions, even around the pole itself, would be subjected to a heat during summer far greater than any which now ever exists in these localities. (Drayson 1873: 148)

Drayson graphically describes the effect of this sudden change from arctic to near tropical conditions:

The natural consequence would be, that the icebergs and ice which had during the severe winter accumulated in high latitudes, would be rapidly thawed by this heat; icebergs far exceeding in number those now annually liberated by the summer sun would be freed from even round the pole, and would float to more southerly regions; streams that had been frozen and blocked with masses of ice would be thawed; whilst the accumulated snow, ice, and frozen streams being thus rapidly and completely thawed, would produce floods which would inundate the country in all directions. A rapid thaw occurring in the present day after a few days' accumulation of snow and frost produces floods: what the effects would be of an almost tropical heating coming suddenly on the frozen masses and glaciers of northern Lapland we can easily imagine. (Drayson 1873: 148–9)

According to Drayson, when the obliquity is around 35°, the seasonal swing of climate and its effects are as follows:

during each year there would have been a fierce oscillation of climate in both northern and southern hemispheres; each winter England and all localities with the same latitude would have experienced an arctic cold, during which the earth's surface would have been covered with a mantle of snow and ice. Then followed the summer, during which excessive heat visited the same localities, liberating icebergs, which would be floated over the country, carrying with them their freight of boulders, gravel, &c. The snow and frozen masses on the mountains would be loosened, and then these masses slipping over the rocks, would leave their scratchings and marks in localities where snow and ice are now rarely found in any quantities. Each winter the whole northern and southern hemispheres would be one mass of ice; each summer nearly the whole of the ice of each hemisphere would be melted and dispersed. These effects would neither be local nor partial; they would occur over the whole northern and southern hemispheres, and would be particularly marked in those latitudes which were then on the borders of the arctic circle—viz. from about 45° to 60° of latitude. (Drayson 1873: 149–50)

The geologist Thomas Belt (1874) believed that Drayson was wrong in assuming that ice-sheets would grow and decay annually. Rather, he suggested that increased obliquity would lead to greater evaporation and thus precipitation, which in turn would result in a year by year accumulation of snow in arctic regions. Under these conditions, the proposed floods would be of far smaller magnitude.

Floods during a long tilt cycle

As recently as 1972, George Williams proposed that the tilt of the Earth's axis changes very, very slowly through 360° over a period of 2500 million years. The climate of the Earth would change drastically through such a long tilt cycle. When the obliquity of the ecliptic is either 0° or 180°, which happens once every 1250 million years, there are no polar ice-caps and a warm, seasonless climate extends to high latitudes. As the obliquity of the ecliptic increases so polar ice-caps form and extend girdles of glacial deposition outwards towards the equator. When the obliquity of the ecliptic has reached either 90° or 270°, which will also occur every 1250 years, the northern and southern hemispheres alternate between a summer of continuous heat and a winter of extreme cold, with the climate at the equator during the equinoxes being similar to the equatorial climate of today. When the rotation stood at an angle which brought it almost into the same plane as the ecliptic, then the climate would be conducive to seasonal floods, similar in kind to those envisaged by Drayson, but much bigger in size. In effect, the floods would be produced by the melting of snow and ice accumulated during the bitterly cold winters.

In support of his simple but elegant hypothesis, Williams interprets the nature of the Late Ordovician glaciation in northwest Africa, as worked out by Fairbridge (1970*a*, 1970*b*, 1971), in terms of a markedly different obliquity to the present one. Features of palaeoclimatic significance described by Fairbridge are: the predominance of sand throughout the section; vast outwash sheets covering thousands of square kilometres with sandstone units up to 20 m thick and hundreds of kilometres in extent showing hydrodynamic evidence of very high velocity currents and suggesting 'a catastrophic decantation of meltwaters' (Fairbridge 1970*a*: 878); evidence for the grounding of ice in fossiliferous marine sediments; the presence of fossils or tracks of trilobites and traces of other marine life, systematically almost through the entire section; evidence for the freezing of loose sands into temporary 'bedrock' upon which tills were deposited; long, parallel grooves cut into outwash sandstones and extending for hundreds of kilometres; and a basement of Precambrian rocks weathered and bleached to depths of 3 to 4 m and capped by a residual hematitic crust or palaeosol. These sedimentary features suggest to Williams severe winters when unconsolidated sands were frozen and, with marine sediments, overridden by advancing ice; and summers when rapid ice melt led to catastrophic floods, marine life spread far into polar waters, weathering produced sand in abundance and affected the underlying rocks, and large blocks of ice might have calved from the melting ice-sheets and slid down the outwash plains to score the frozen sands for many kilometres.

Partial floods as grand debacles

It is perhaps not surprising that the notion of a universal Flood was anathema to Lyell. He notes in his *Principles* that the matter of whether the deluge of the Scriptures was universal in reference to the whole surface of the globe, or only so with respect to that portion of the globe inhabited by men, had been debated for many centuries. Certainly, the universality of the Noachian Flood was a problem which had exercised the best minds of the Renaissance. Other than the authority of Holy Writ, the best evidence for global flooding was the wide spread of fossils, which most scholars agreed could be attributed to the Flood. But if the Flood was universal, how could the ferocious beasts in North America, so unlike the Old World animals, be explained? Why had not the Flood drowned them, too? And what could be the source of enough water to cover the loftiest mountains? In the face of these vexing matters, some writers argued in favour of the idea that the Flood had been partial, rather than universal; it had inundated a large part of the Middle East, but had not submerged the entire globe.

One of the first scholars to ponder the general nature of the Flood was Leonardo da Vinci. With his customary penetrating logic, da Vinci questioned whether Noah's Flood could have been universal, and argued against its universality on the following grounds:

We have it in the Bible that the said Flood was caused by forty days and forty nights of continuous and universal rain, and that this rain rose ten cubits above the highest mountains in the world. But consequently if it had been the case that the rain was universal it would have formed in itself a covering around our globe which is spherical in shape; and a sphere has every part of its circumference equally distant from its centre, and therefore on the sphere of water finding itself in the aforesaid condition, it becomes impossible for the water on its surface to move, since water does not move of its own accord unless to descend. How then did the waters of so great a Flood depart if it is proved that they had no power of motion? If it departed, how did it move, unless it went upwards? At this point natural causes fail us, and therefore in order to resolve such a doubt we must needs either call in a miracle to our aid or else say that all this water was evaporated by the heat of the sun. (da Vinci 1977: i. 300)

Another early writer to challenge the universality of the Flood was the French intellectual and diplomat, Isaac de la Peyrère (1595–1676). In his *Praeadamitae* (1655), la Peyrère questioned the ability of animals to migrate from Mount Ararat to all other parts of the world. He suggested that the Deluge had been a local event confined to Europe and the Middle East, and that it did not destroy the birds, beasts, and plants living outside that region. For sporting such heretical views, la Peyrère was condemned by the parliament in Paris. He fled to Brussels but was arrested there, and then escorted to Rome to sign a public retraction of his views in the presence of Pope Alexander VII. But la Peyrère was not the only scholar to espouse a partial flood hypothesis: Isaac Vossius

(1618–85), and Edward Stillingfleet, Bishop of Worcester (1635–99) also considered the possibility that the Noachian Flood might have been a local event. Vossius presented his views in his *Dissertatio de vera aetate mundi* (1659), and Stillingfleet his in his *Origines sacrae* (1662, 1836). In the eighteenth century, Richard Clayton, Bishop of Clogher (1695–1758), supported the notion of a partial Flood in his *A vindication of the histories of the Old and New Testament in answer to the objections of the late Lord Bolingbroke*, published in Dublin in 1752. It was Clayton's views on the Flood which spurred Alexander Catcott into devising a system of Earth history which explained the manner by which a global flood might be produced.

Lyell, not wishing to outrage his popular reading public by overtly rejecting Scriptural teachings, tactfully argues that, if the interpretation of the Flood as a local event be admissible, then there are two classes of phenomena in the configuration of the Earth's surface which might account for such an event. These phenomena are, firstly, extensive lakes standing above the level of the sea; and, secondly, large tracts of land lying below sea level. Lyell explains the significance of such elevated and depressed regions to his theory of local flooding:

When there is an immense lake, having its surface, like Lake Superior, raised six hundred feet above the level of the sea, the waters may be suddenly let loose by the rending or sinking down of the barrier during earthquakes, and hereby a region as extensive as the valley of the Mississippi, inhabited by a population of several millions, might be deluged. On the other hand, there may be a country placed beneath the mean level of the ocean, as was shown to be the case with part of Asia, and such a region must be entirely laid under water, should the tract which separates it from the ocean be fissured or depressed to a certain depth. The great cavity of western Asia is eighteen thousand square leagues in area, and is occupied by a considerable population. The lowest parts, surrounding the Caspian Sea, are about 350 feet below the level of the Euxine,—here, therefore, the diluvial waters might overflow the summits of hills rising 350 feet above the level of the plain; and if depression still more profound existed at any former time in Asia, the tops of still loftier mountains may have been covered by a flood. (Lyell 1834: iv. 146–7)

Floods and lake bursts

It was a view widely held among early nineteenth-century geologists that the Earth's surface had formerly been spotted with a galaxy of lakes. At times, the barriers which impounded the lake waters were brought down by catastrophic earthquakes. The water in a lake then burst through the ruptured dam and rushed down the river valley as a violent debacle. Geologists who subscribed to this view saw debacles as important agents in the moulding of the Earth's surface features.

The efficacy of debacles in eroding valleys was well known by mineral prospectors. A common technique used in prospecting was 'hushing'. This involved damming a stream to form a pond; then, when the pond was big enough, allowing it to sweep away the dam. The resulting surge of water, as it rushed down the valley, stripped the soil and mantle from the valley sides. The eroded area, or 'hush-gutter', was then inspected for signs of mineral veins in the freshly exposed outcrop of bedrock. The same process occurred naturally, but with tragic consequences, in the Val de Bagnes, Switzerland, in April 1818. An avalanche of ice from the Getroz glacier cascaded into the valley above Sembrancher, blocking the course of the river Dranse. A lake of some eight hundred million cubic feet of water formed behind the blockage. In order to draw off some of the impounded water, a tunnel was dug through the upper part of the dam. But, in mid-June, the dam suddenly gave way. The remaining water surged towards Lake Geneva, devastating the valley and killing many people. The tragic event was reported in detail by Basil Hall in his book entitled *Patchwork* (1841). He tells us how

the church presented one of its corners to the advancing tide, but although it escaped destruction, it was nearly half full of sand, mud and stones, brought there by the flood. The pulpit just peeped above the mass of rubbish, but the altar was no longer visible, being quite buried under the mud; . . . all the hedges, garden-walls, and other boundary lines and landmarks of every description were obliterated under one uniform mass of detritus, which had levelled all distinctions in a truly sweeping and democratic confusion. In every house, without exception, there lay a stratum of alluvial matter several feet in thickness, so deposited that passages were obliged to be cut through it along the streets, as we see roads cut in the snow after a storm. On the side of the building, facing up the valley, was collected a pile of large stones, under all these a layer of trees, with their tattered branches lying one way, and their roots the other. Next came a network of timber beams of houses, broken doors, fragments of mill-wheels, shafts of carts, handles of ploughs, and all the wreck and ruin of the numerous villages which the debacle had first torn to pieces, and then swept down the valley in one undistinguishable mass. . . . From every house and from behind every tree there extended down the valley a long *tail* or train of diluvial rubbish, deposited in the swirl, or, as a sailor would say, in the eddy, under the lee of these obstacles. All over the plain, large boulders or erratic blocks lay thickly strewed. These varied in size from a yard to a couple of yards in diameter; but just at the point where the ravine of the Drance leaves the mountains, and joins the open valley of Martigny, I examined some enormously large masses of granite, which the inhabitants assured me had been brought down and placed there by the sheer force of the debacle. . . . I well remember the awe and wonder with which I looked at one of the masses of rock pointed out to me, which the stream had evidently projected fairly out of the gorge into the plain. It measured 27 paces round, 12 feet in height, and 12 feet across in one direction, which I fixed upon as about the average. It was a rude pyramidal shape. Further up the glen, I came to many rocks, which, though much larger than the one I mentioned, bore indubitable marks of having been in motion. (Quoted in Howorth 1893: ii. 874–5)

This remarkable debacle made a great impression on geologists, some deeming it to be a fine example of a kind of event common throughout the history of the Earth.

Debacles in North America

The notion of debacles was very popular among North American geologists during the nineteenth century. It was a widely held view that a great barrier had once held back a vast inland sea in the area of the Great Lakes. The lake waters eventually broke through the barrier to produce a debacle on a grand scale. Samuel Akerly, in a geological account of Duchess County, New York, invoked this grand debacle to explain the 'alluvion of sand, stone and rocks' in the southern part of the region:

After the waters of the Deluge had retired from this continent, they left a vast chain of lakes, some of which are still confined within their rocky barriers; others have since broken their bounds, and united with the ocean. The highlands of New-York was the southern boundary of a huge collection of water, which was confined on the west by the Shawangunk and Kaats-kill mountains. The hills on the east of the Hudson confined it there. When the hills were cleft and the mountains torn asunder, the water found vent and overflowed the country to the south. . . . The earth, sand, stones and rocks brought down by this torrent were deposited in various places: as on this (New-York) island, Long Island, Staten Island and the Jerseys'. (Quoted in Chorley *et al.* 1964: 239)

In his 'Observations on the geology of North America' (1818), Samuel L. Mitchill suggested that the Great Lakes were a shrunken remnant of a great internal sea, the waters of which were impounded by a great ridge of land that seemed to have

circumscribed the waters of the original Lake Ontario and to be still traceable as a mountainous ridge beyond the St. Lawrence in upper Canada, passing thence into New York, where it formed the divide between the present lake and the St. Lawrence and continued to the north end of Lake George, apparently crossing the Hudson above Hadley Falls. Thence he believed it to run toward the eastern sources of the Susquehanna, which it crossed to the north of Harrisburg, and continued in a southeasterly direction until it entered Maryland, passing the Potomac at Harpers Ferry into Virginia, where it became confounded with the Alleghany Mountains. (Merrill 1924: 617)

Eventually the ridge gave way at various points,

the pent-up waters rushing through and carrying devastation before them like the waters from cloudburst or bursting reservoirs of today, but on a thousand-fold larger scale. By this bursting all the country on both the Canadian and Fredonian sides must have been drained and left bare, exposing to view the water-worn pebbles, and the whole exhibition of organic remains there formed. Great masses of primitive rocks from the demolished dam, and vast quantities of sand, mud, and gravel were carried down the stream to form the curious admixture of primitive with alluvial materials in the regions below. (Merrill 1924: 617)

William H. Keating, a member of Major S. H. Long's 1823 expedition to the Great Lakes and sources of the Mississippi, is reported by Merrill to have made the following observations in the area of the extinct glacial Lake Agassiz:

The whole region comprising the headwaters of the Winnipeek River was looked upon as having been at a comparatively recent period an immense lake interspersed with innumerable barren, rocky islands, which had been drained by the bursting of the barriers which tided back the waters.

The innumerable bowlders which he found covering the valley were regarded as due to the flood of waters caused by the bursting of these natural dams. (Merrill 1924: 104)

A grand debacle was also identified as a possible cause of the original valley and falls of Niagara. Henry H. Rogers writes:

It is a very generally received opinion and may so far as present evidence extends, be taken for granted, that the country adjacent to Niagara and the lakes was originally covered with a vast lake, or rather inland sea, which some change in the configuration of the region contracted to the still very extensive masses of fresh water now remaining. The passage of such a body of water over the surface would deeply indent all the exposed portions of the land. Rushing in its descent from Lake Erie to Lake Ontario, from a higher to a lower plain, and across a slope like that at Queenstown, it would inevitably leave a deep and long ravine. But further, the whole of this region has been grooved and scarified by the same far sweeping currents which denuded the entire surface of North America, and strewed its plains and mountains with boulders, gravel and soil from the north. Such a diluvial valley, of greater or less length and depth was, I cannot help believing, probably the commencement of the present remarkable trough below the Falls. (Rogers 1835: 329)

While the role of a grand debacle was not generally questioned, the detail of its effects on surface topography was queried. Thus, J. B. Gibson took Rogers to task on a number of particulars:

When we see the river [the Niagara] working in the rock like an endless saw, it is difficult to think that it did not make the groove in which we find it. If this groove were originally but a valley of denudation, why are its sides perpendicular even at the brink, and why is the original inclination of its slope broken by a cataract now? In the opinion of Professor Rogers and many others, an inland sea, vastly more immense than the present fresh water lakes, sent a current along the course of the Niagara river, tearing up the exposed portion of the land, and imperfectly excavating the rough and unshapen trough below the falls. The traces of an overwhelming current are doubtless every where visible; and it is reasonable to suppose that, seeking the lowest part of the barrier, it would gradually narrow and confine its action at that point, at least sufficiently to mark out the course of the subsequently diminished stream. But we are unable to imagine how a wide spread torrent could have spent its entire action on a strip six hundred yards in breadth, giving to the sides of the gutter made by it, the character and appearance of perpendicular walls. No such walls are found in the water gaps of the Alleghany mountains. (Gibson 1836: 204–5)

The notion of debacles went west with the geologist-explorers during the middle of the nineteenth century. John Strong Newberry (1822–92) was the surgeon and naturalist on the first truly scientific expeditions into the west of North America. In 1857, he joined an expedition to the lower Colorado river. Having run aground in the steamboat *Explorer*, the party was forced to strike out overland, and it entered the Grand Canyon on foot, from the south side at Diamond Creek. To explain how the Grand Canyon and its tributary valleys were formed, Newberry suggested that the river Colorado had been impounded by mountain ranges in a series of basins. On spilling over the barrier, the water, released from its impoundment, gouged out the canyons (Newberry 1861: 47).

A uniformitarian view of debacles

One of the first writers to attribute valley formation to the action of debacles was the Italian traveller, Giovanni Targioni-Tozzetti (1712–84). From his excursions in Tuscany, Targioni-Tozzetti concluded that the fossil bones of elephants and other quadrupeds found in the valleys of the Arno, Val di Chaina, and Ombrosa belonged to beasts which had once dwelt in the valleys, and had not been transported there by Hannibal, or the Romans, or the Noachian Deluge. Nor had the valleys themselves been sculpted by the Flood: they were excavated, after the retreat of the ocean, by rivers and by floods, the floods being caused by the bursting of barrier lakes (Lyell 1834: i. 69–70; von Zittel 1901: 34–5).

As has already been mentioned, Lyell explained how the release of water from dammed lakes might produce features in the landscape which could be mistaken for signs of the universal Noachian Flood. Another firm believer in the efficacy of debacles as an erosive tool was Henry de la Beche (1834). In his *The geological observer* (1851), he discussed the various types of dam which might be thrown across the course of a river to impound a lake and cause an eventual debacle. The dams might be made of ice, lava, landslide materials, or a large alluvial fan. But by 1851, there were few geologists who still attached much attention to the importance of debacles as agents of erosion (Davies 1969: 248). It was not until the 1920s that the notion of debacles was again taken seriously.

The Spokane Flood

The biggest debacle for which there is convincing evidence took place between 13 000 and 18 000 years ago in south-eastern Washington state, and involved two outbursts from Glacial Lake Missoula following the failure of impounding dams of ice (Baker 1978) (Figure 8.3). The evidence of its occurrence was first recognized by J. Harlen Bretz (1923*a*, 1923*b*). By assiduous field observation and mapping, Bretz revealed 'a pattern of abandoned erosional waterways, many of them streamless canyons (coulees) with former cataract cliffs and plunge basins, potholes and deep rock basins, all eroded in the underlying basalt of the gently southwestward dipping slope of that part of the Columbia Plateau' (Bretz 1978: 1). In accounting for such a pattern of landscape features, Bretz suggested that a debacle had been the cause, which he later designated the Spokane Flood. This brief but immense

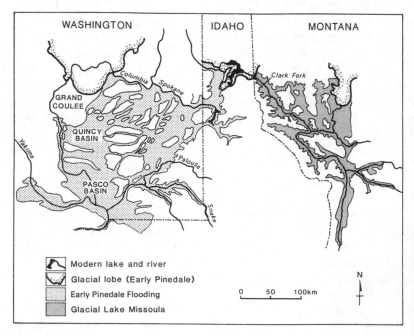

Fig. 8.3. Glacial Lake Missoula and the Channeled Scabland. (After Baker 1978)

flood had filled normal valleys to the brim, and had then spilled over the former divides, eroding the summits to complete the network of drainage ways. To describe such a complex landscape as a valley seemed to Bretz inadequate. Instead, 'the abandoned rock-bound former waterways were called channels, and the entire composite area was named "Channeled Scabland"' (Bretz 1978: 1). Here is Bretz's description of the Channeled Scabland, as reported in the *Geographical Review*:

No one with an eye for landforms can cross eastern Washington in daylight without encountering and being impressed by the 'scabland'. Like great scars marring the otherwise fair face of the plateau are these elongated tracts of bare, or nearly bare, black rock carved into mazes of buttes and canyons. Everybody on the plateau knows scabland. It interrupts the wheat lands, parceling them out into hill tracts less than 40 acres to more than 40 square miles in extent. One can neither reach them nor depart from them without crossing some part of the ramifying scabland. Aside from affording a scanty pasturage, scabland is almost without value. The popular name is an expressive metaphor. The scablands are wounds only partially healed—great wounds in the epidermis of soil with which Nature protects the underlying rock.

   With eyes only a few feet above the ground the observer today must travel back and forth repeatedly and must record his observations mentally, photographically, by sketch and by map before he can form anything approaching a complete picture. Yet long before the paper bearing these words has yellowed, the average observer, looking down from the air as he crosses the region, will see almost at a glance the picture here drawn by piecing together the ground-level observations of months of work. The region is unique: let the observer take the wings of the morning to the uttermost parts of the earth: he will nowhere find its likeness.

   Conceive of a roughly rectangular area of about 12,000 square miles, which has been tilted up along its northern side and eastern end to produce a regional slope approximately 20 feet to the mile. Consider this slope as the warped surface of a thick, resistant formation, over which lies a cover of unconsolidated materials a few feet to 250 feet thick. A slightly irregular dendritic drainage pattern in maturity has been developed in the weaker materials, but only the major stream ways have been eroded into the resistant underlying bed rock. Deep canyons bound the rectangle on the north, west, and south, the two master streams which occupy them converging and joining near the southwestern corner where the downwarping of the region is greatest.

   Conceive now that this drainage system of the gently tilted region is entered by glacial waters along more than a hundred miles of its northern high border. The volume of the invading water much exceeds the capacity of the existing stream ways. The valleys entered become river channels, they brim over into neighbouring ones, and minor divides within the system are crossed in hundreds of places. Many of these divides are trenched to the level of the preexisting valley floors, others have the weaker superjacent formations entirely swept off for many miles. All told, 2800 square miles of the region are scoured clean into the basalt bedrock, and 900 square miles are buried in the debris deposited by these great rivers. The topographic features produced during this episode are wholly river-bottom forms or are compounded of river-bottom modifications of the invaded and over-swept drainage network of hills and valleys. Hundreds of cataract ledges, of basins and canyons eroded into bed rock, of isolated buttes of the bed rock, of gravel bars piled high above valley floors, and of island hills of the weaker overlying formations are left at the cessation of this episode. No fluviatile plains are formed, no lacustrine flats are deposited, almost no debris is brought into the region with the invading waters. Everywhere the record is of extraordinarily vigorous sub-fluvial action. The physiographic expression of the region is without parallel; it is unique, this channeled scabland of the Columbia Plateau. (Bretz 1928: 446)

The key to understanding the physiography of this unique region lay in explaining how to produce enough glacial meltwater to do the job. In the event, Bretz called for a grand debacle—the Spokane Flood. This suggestion generated a flood of high-handed criticism almost as big as the Spokane Flood itself. Here is how Bretz recalled the episode in later years:

Catastrophism had virtually vanished from geological thinking when Hutton's concept of 'the Present is the key to the Past' was accepted and Uniformitarianism was born. Was not this debacle that had been deduced from the Channeled Scabland simply a return, a retreat to catastrophism, to the dark ages of geology? It could not, it must not be tolerated.

This, the writer of the 1923 article learned when, in 1927, he was invited to lecture on his finding and thinkings before the Geological Society of Washington, D.C. an organization heavily manned by the staff of the United States Geological Survey. A discussion followed the lecture, and six elders spoke their prepared rebuttals. They demanded, in effect, a return to sanity and Uniformitarianism. (Bretz 1978: 1)

But Bretz stood by his guns and doggedly pursued his research into this enormous debacle:

The upstart theorist was not upset nor silenced. Despite his knowledge that the country was full of other dissenters to his flood theory, he proceeded to publish more papers on his favorite topic, now named the Spokane Flood. He described other features of the afflicted plateau which he claimed were inexplicable without his flood of glacially derived meltwater. His apostacy would not be corrected as advised by the elders. The one-man rebellion was still alive. (Bretz 1978: 1)

Bretz painstakingly brought to light more and more detail of the flood and its effects. He managed to trace the flood down the Columbia river as far as Portland, Oregon, adding a 200-square-mile delta in the Willamette Valley. His prize discovery—the source of the voluminous flood waters—was reported in 1930:

Clark Fork of the Columbia River, draining a large mountainous region of western Montana, had been dammed by the Cordilleran ice sheet at its traverse of the Idaho panhandle. This formed an immense glacial lake with an estimated volume of 500 cubic miles. The lake had been named some years before as Glacial Lake Missoula. The first geologist to describe the lake ironically was one of the six challenging elders in Washington in 1927 and the author of a short paper on problematical features, perhaps glacial in origin, in what came to be known as the Channeled Scabland.

If Lake Missoula had a properly located place for its ice dam and a clear route thence to the Channeled Scabland, then presto, we would have the big problem solved. Missoula's depth at the dam was known from its shorelines to have been 2000 feet, and there was a clear route to Spokane and the Scabland. A catastrophic failure of the dam would release 500 cubic miles of glacially derived water with adequate gradient to Spokane. (Bretz 1978: 1–2)

Bretz had to wait many years till his outrageous hypothesis, for so it was regarded, was vindicated. In 1942, J. T. Pardee accepted that Lake Missoula was the source of flood waters which had rushed through the Channeled Scabland. But it was not until 1956, with the publication of a report on a further set of field investigations, that the sharp knives of the critics were finally turned. In a field study made in the summer of 1952, Bretz, approaching seventy years of age, discovered a criterion of undeniable validity for the occurrence of a flood:

Hidden largely by sagebrush were numerous occurrences of current ripple marks. They were discovered because the U.S. Bureau of Reclamation had taken aerial photographs of the area to be irrigated with Grand

Coulee water. Then it became clear that some gravel surfaces, curiously humpy, were covered with giant current ripples. An investigator, standing between two humps, could not see over either one. Indeed, the size of these ripple ridges made them really small hills. Finally came the discovery of giant current ripples in parts of Lake Missoula where, in a catastrophic emptying, strong currents were formed. (Bretz 1978: 2; see Bretz *et al.* 1956)

And in 1973, Victor Baker, by measuring records for depths of water and water-surface gradients in channels with proper cross sections, was able to estimate the discharge of water during the flood. The flood discharge reached 21.3 million m^3/sec, and in some channels the flood flow velocity touched 30 m/sec; but even at that phenomenal discharge, it would take a day to empty the lake of its 2.0×10^{12} m^3 of water (Baker 1973). Further studies have showed that at least five major cataclysmic floods occurred during the Quaternary in the general vicinity of the Channeled Scabland, of which the Spokane Flood was the last.

The Bonneville Flood

Bretz may be correct in asserting that the Spokane Flood beggars parallel, but the Lake Bonneville Flood comes a close second. This debacle occurred about 15 000 years ago when Pleistocene Lake Bonneville overtopped its rim at Red Rock Pass in south-eastern Idaho and rapidly lowered, decanting about 4700 cubic kilometres of water down the Snake River. This debacle was studied by Malde (1968), who traced it through the Snake River Plain of southern Idaho to Hells Canyon. In rushing down the Snake River, the flood caused extensive erosion and deposition. Today, the valley displays impressive abandoned channels, areas of scabland, and gravel bars composed of sand and angular and rounded boulders up to 3 m in diameter. Using a step-backwater computational technique, Jarrett and Malde (1987) have shown that the peak discharge for the flood through the constricted reach of the Snake River Canyon at the mouth of Sinker Creek was between 793 000 and 1 020 000 m^3/sec. They estimate that at this rate of discharge, the shear stress for the flood would have been 2500 N/m^2, and the unit stream power would have been 75 000 N/m/sec. This compares with shear stress and unit stream power for recent floods of the Mississippi and Amazon rivers of 6 to 10 N/m^2 and 12 N/m/sec.

Other causes of big floods

Massive floods, though not as powerful as the Spokane and Bonneville debacles, can be produced by extreme precipitation events. These 'superfloods' have confounded hydrologists because they cannot be understood in terms of drainage basin hydrology; rather, they result from anomalies in the atmospheric circulation on an almost hemispherical scale (Baker 1983). Although these superfloods are short-lived, they are noteworthy for their long-term effects on the landscape (Baker 1977). There is probably a limit on the discharge of these floods, owing to physical constraints imposed by a particular drainage basin. Certainly, Partridge and Baker (1987) found this to be the case for events with a recurrence interval of up to 2000 years at the Salt River watershed, Arizona.

Interesting as these 'superfloods' are, they are insignificant in comparison with the scale of flooding hypothesized to arise from the impact in the ocean of a meteorite, or from the fast tumble of the Earth. The gigantic waves produced by both fast tumble and bombardment would produce floods which could truly be prefixed by the word 'super'.

9

Catastrophic Neodiluvialism
Floods, fast tumble, and meteorite bombardment

In the previous chapters, two natural events have been identified which may cause superfloods of relatively limited extent: the sudden release of water from dammed lakes; and periods of exceptionally high precipitation occurring in conjunction with hemispherical anomalies in the atmosphere. Two other natural events are capable of causing floods of truly super magnitude: fast geographical or astronomical pole shift; and the impact of an asteroid or comet in the ocean. The first event, fast Earth tumble or fast tilt changes, would produce grand cataclysms. The second event, cometary or asteroidal impact, could produce superflooding of continental lowlands and possibly highlands.

It is now generally accepted that the poles do wander gradually over long periods of time in response to inequalities in the distribution of terrestrial mass. However, the consensus among scientists who are broadminded enough not to dismiss it out of hand, is that fast pole shift is very unlikely to occur, and the fast pole shifters are usually regarded with scepticism, if not downright hostility. It seems only fair, however, in a book dealing with the history of ideas concerning diluvialism, that the unconventional views of these independent thinkers, as Patrick Moore has politely called such eccentric scientists, are mentioned, albeit briefly. But, although both fast pole shift and bombardment by asteroids and comets are claimed to produce grand cataclysms, and either could therefore form the basis of a neodiluvialist system of Earth history, bombardment will form the basis of the neodiluvialism proposed latter in the chapter as it is at present a far more convincing hypothesis than the rather implausible notions of fast pole shift.

Cataclysms and fast pole shift

There is no doubt that were the Earth to tumble fast, the consequences would be catastrophic. The terms used by the advocates of fast tumble read like a set of key words compiled for a book penned by a catastrophist of the old school: they include tidal waves, shock waves, hurricanes, changes of sea level, mass extinction of organisms, climatic change, episodes of widespread and intensive volcanism, and the cataclysmic change of landforms. Violent and sudden processes may no longer be dispensed by the hand of God, but, according to these latter-day catastrophists, their power is still awesome. However, it would be wrong to surmise that all catastrophic processes wreak widespread destruction: less violent, and moderately sudden processes may induce more subtle changes in Earth systems, such as a fairly rapid, but not necessarily destructive, change of climate.

Cataclysms and an unbalanced Earth

The chief hypotheses concerning a rapid tumble of the Earth involve imbalances in terrestrial mass and gravitational interaction with stray planets. The first of these possibilities is certainly more acceptable to Earth scientists than the second. It was first suggested by Hugh Auchincloss Brown. Born in 1879, Brown graduated from Columbia University in 1900 with an engineering degree. In 1911 he became interested by reports of mammoths found frozen in the Arctic in a life-like condition with grass still in their mouths. He set about constructing an hypothesis, the gist of which is that the accumulation of ice at one or both of the poles periodically upsets the equilibrium of the spinning Earth causing it to turn sideways or tumble (Brown uses the word careen), in the manner of an overloaded canoe, bringing the ice-caps near to equatorial positions. As the continents tumble against the oceans, water would rush over the continents to produce a cataclysm with awesome destructive power: vegetation would be crushed to pulp and animals would be obliterated. Brown claims that his hypothesis might explain, among many other things, the various flood myths and the presence of quick-frozen mammoths, and other large herbivores, in the Arctic.

Brown published his ideas privately in 1948 in a treatise entitled *Popular awakening concerning the impending flood* and as a book, *Cataclysms of the Earth*, in 1967. He died in 1975 at the age of ninety-six, still firmly holding his belief that the Earth was soon to suffer a cataclysm which would leave New York under water. His work has largely been ignored by the geological establishment, the reception of his 1948 treatise being for the most part scornful. Walter Sullivan (1974) paid lip-service to Brown's ideas, but argued that the stability of the Earth's spin is too great for it to change suddenly, even if a lopsided chunk of ice should accumulate in the southern polar region. White (1986) believed that he had found a fatal flaw in Brown's hypothesis: the Antarctic ice-cap may be as much as 20 million years old, and therefore the Earth cannot have tumbled through 80° during that time. However, a fossil forest has recent been discovered in Antarctica which is only 2 to 3 million years old (Anderson 1986), a fact which seems to deflate White's objection. In fact the most controversial, indeed ridiculous, point Brown makes is that the tumbling Earth actually slips through the equatorial bulge to produce a ground wave some 21 km high which appears to an observer standing between the equator and the pole to travel at hundreds of kilometres per hour. If the Earth's crust had been subjected to this process in the recent or distant past, then the field evidence of such a catastrophe would surely be unequivocal, though it is doubtful if any observers on the Earth, of any species, would be around to interpret it!

Summing up Brown's work, White (1986: 85) writes that his argument is not without merit, if not fully acceptable; that his concept of a tumbling Earth caused by an off-centred mass of ice has an appealing elegance in the light of the enigma of quick-frozen mammoths; that he has gathered data and offered what seems to be the beginning of a solution to a global mystery; but that major, more plausible modifications and alternatives to Brown exist. In conclusion, 'It seems fairest to say aloud on Brown's behalf, as Galileo is supposed to have muttered after being forced by the Inquisition to recant his idea that the earth revolves around the sun, "Nevertheless, it moves"' (White 1986: 85).

Building on the work of Brown, Charles Hapgood, in his book *Earth's shifting crust*, proposed that, instead of the Earth as a whole tumbling during a pole shift, only an outer shell moves. He first considered Brown's two basic assumptions concerning the centrifugal effect of bodies rotating off-centre and the stabilizing effect of the equatorial bulge, and found them both to be sound. He then sought the ratio of the unstabilizing centrifugal effect of an ice-sheet to the stabilizing effect of the equatorial bulge. It became clear that

the force of the ice cap would either have to overcome the total stabilizing centrifugal effect of the bulge, or it would have to shatter the crust, so that the Earth could start to rotate farther off centre, thereby initiating a chain reaction of increasing centrifugal effects. (Hapgood 1958: 17)

In trying to estimate the centrifugal effect of the ice-sheet, Hapgood noticed that the eccentricity related to the Chandler wobble recognized by Brown was small compared with the eccentricity of the Antarctic ice-sheet itself, the centre of gravity of which lies about 550 km from the rotation axis.

Next Hapgood, with the help of the United States Coast and Geodetic Survey, measured the stabilizing effect of the equatorial bulge and discovered that it was several thousand times greater than the centrifugal effect of the off-centred polar ice-sheet. His investigation would have ceased at that point had not his friend James Hunter Campbell, an engineer who had helped develop the Sperry gyroscopic compass, suggested that if the ice-sheet did not have enough force to tumble the whole planet, it might have sufficient force to displace the Earth's crust over the underlying layers. The outcome of Hapgood's collaboration with Campbell was, as Albert Einstein wrote in the introduction to Hapgood's (1958) book, an original hypothesis of great simplicity and great importance. Briefly stated, Hapgood and Campbell's hypothesis is that, periodically, ice at one of the poles accumulates so fast that the rotation axis cannot adjust to the off-centred, non-isostatically compensated ice mass and the Earth's crust slips to compensate. In passing over the equatorial bulge the crust is deformed, producing stresses of the order of 100 million dynes/cm^2. The rate of crustal slipping slows as the ice-sheet melts on reaching warmer latitudes, the entire slipping episode taking something in the order of 10 000 years.

A number of objections to aspects of the original Hapgood–Campbell hypothesis have been voiced. For instance, Chadwick (1962) makes three points. Firstly, recent gravity and seismic measurements made in Antarctica suggest that the ice-cap is, if anything, over-compensated isostatically. Secondly, without a proper understanding of the mechanical properties of the crust and underlying layers, it is difficult to say whether a tangential force would displace the whole crust or whether it would merely lead to local crustal deformation. And thirdly, for crustal slipping to occur repeatedly in the manner envisaged by Hapgood and Campbell, eccentric ice masses must be common throughout geological time, a requirement which is at odds with the geological evidence. Asimov (1979: 163–4) is emphatic that crustal slip has never occurred. He asserts that if it did take place, then the crust would have to tear apart as it slipped from polar regions to equatorial regions, and crumple together as it moved from equatorial regions to polar regions, the extreme degree of tearing and crumpling involved leaving clear signs in the crust, signs which nobody has yet found.

In 1970 Hapgood, in the light of new studies of the nature of the Earth's crust which indicated that the forces responsible for crustal shifts lie at some depth rather than on the surface, abandoned his contention that ice-sheets are the trigger mechanism for shifts of the crust. However, the main thrust of his argument, that the occurrence of crustal displacements is a common geological event which has been responsible for the formation of many of the Earth's features, remained unaltered. Hapgood (1970) identified over 200 episodes of crustal displacement during the Phanerozoic. Each episode took several thousand years and involved an Earth tumble of about 30°. The average rate of crustal displacement is thus of the order of a kilometre per year, a figure far lower than the rates proposed by Brown, who envisages a nearly instantaneous shift of almost 90°. Hapgood admits that he cannot identify a specific cause for displacements, but he does indicate the general direction from which a cause may be found: gravitational imbalances either within the lithosphere or immediately below it. Such imbalances do, without question, exist. Whether they are big enough to cause a fast pole shift is debatable (White 1986: 96).

A test of fast tumble

In a pioneering study, Edward Weyer (1978) attempted to make an empirical evaluation of fast tumble hypotheses which invoke mass redistribution at the Earth's surface, particularly the growth and decay of ice-sheets, as a causal mechanism. He studied the heights of 48 fossil shorelines from widely scattered regions dated at between 14 700 and 28 000 years old to see if the surface load of ice during the last glacial episode had caused detectable changes in the position of the geographical poles. If the sea level had changed owing to the removal of water from the oceans to feed the growing ice-sheets, then shorelines of a similar age should occur at similar heights, but this is not the case (Figure 9.1). When allowance is made for a shifting pole, then the seemingly contradictory samples congregate along a regular curve (Figure 9.2). The sea-level curve revealed by correcting for pole shift indicates that two glacial advances took place during the period of study, whereas conventional wisdom, based on a sea-level curve which does not take account of pole shift, recognizes one glacial advance. Weyer points out that his curve for polar movement resembles a sine curve with a twice recurring period of about 5600 years. He also observes the slight asymmetry of the polar movement curve suggesting that the ice-sheets grew at a faster rate than they melted. The changes of sea level envisaged by Weyer would lead to a gradual inundation of large areas of continental lowland, producing new shorelines and burying river courses and their associated terraces. Weyer's work is certainly an exciting, empirical avenue of enquiry in a subject otherwise steeped in speculation.

Stray planets

The suggestion made by Immanuel Velikovsky in a series of contentious books (Velikovsky 1950, 1952, 1955), that the planet Venus was during historical times ejected from the centre of Jupiter and in travelling to its present orbit twice passed close by the Earth, is regarded by most astronomers as ridiculous. Velikovsky claimed that the forces unleashed by the close encounters with Venus led to planetary cataclysms which, in the case of the Earth, involved superfast tumbles, rapid changes in the posture of the rotation axis, and a change in the direction of rotation. He predicted that, following a sudden shift of the Earth about its rotation axis, air and water would continue to move owing to inertia. Hurricanes would sweep the Earth and the oceans would rush over the continents. Volcanoes would erupt, setting light to forests. The burning forests would be wrested from the ground in which they grew and be piled up in huge heaps by the hurricanes and wild seas. Many species and genera of animals on land and sea would be destroyed.

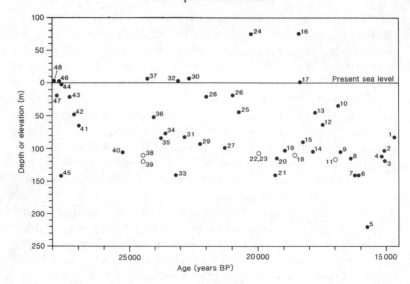

FIG. 9.1. Forty-eight dated shoreline samples from widely scattered regions, plotted with reference to present sea level. Note that the depth and elevations at which the samples were found range erratically over about 300 m, whereas samples with similar dates should follow similar levels. The numbered sites are located as follows (precise locations are given by Weyer 1978): 1. West Africa. 2. West Africa. 3. West coast of Mexico. 4. East China Sea. 5. East China Sea. 6. West Mediterranean. 7. West Mediterranean. 8. West coast of Mexico. 9. West coast of Mexico. 10. Florida. 11. Southern California. 12. South-east Caribbean. 13. South-east Caribbean. 14. West coast of Mexico. 15. Western Mediterranean. 16. Juan Fernandez. 17. Borneo. 18. Southern California. 19. Western Mediterranean. 20. West coast of Mexico. 21. Western Mediterranean. 22. Southern California. 23. Southern California. 24. Juan Fernandez. 25. Florida. 26. Georgia. 27. Western Mediterranean. 28. Florida. 29. Panama. 30. Canary Islands. 31. West Africa. 32. Corsica. 33. East China Sea. 34. Western Mediterranean. 35. Western Mediterranean. 36. Western Mediterranean. 37. Florida. 38. Southern California. 39. Southern California. 40. Western Mediterranean. 41. East China Sea. 42. Western Mediterranean. 43. Texas. 44. Georgia. 45. Western Mediterranean. 46. Florida. 47. Georgia. 48. Florida. (After Weyer 1978)

A slightly less unpalatable version of the Velikovskian model has been put forward by Peter Warlow (1978, 1982). Warlow argues that, in passing near the Earth, an extraterrestrial body of comparable mass to the Earth will exert a gravitational pull which, owing to the equatorial bulge, will create a torque (turning force). The effect of this torque will be to cause the Earth either to tumble a little or to tumble right over so that the geographical poles swap places and the Earth stands upside-down. Warlow suggests that a full reversal of the Earth could take place in a day! So, as in

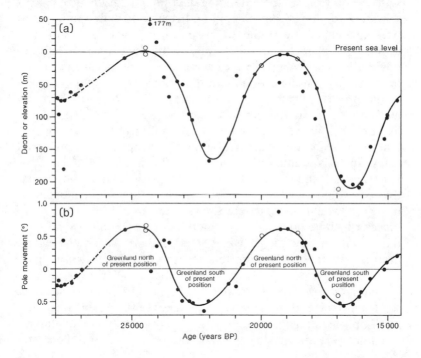

FIG. 9.2. (a) When a shifting pole is postulated, the individual shoreline samples plotted in Figure 9.1 congregate along the curve indicated. (b) The curve plotted shows the pole shift that would transform the seemingly random distribution of shoreline samples in time (Figure 9.1) into the orderly configuration shown in Figure 9.2(a). The polar curve starts down each time ice-sheets begin to grow, and there is a corresponding drop in the sea-level curve, a fact consistent with the theory that the centrifugal force created by ice-sheets would cause the Earth to tumble. (After Weyer 1978)

Velikovsky's hypothesis, superfast rates of tumble are envisaged. Like Velikovsky, Warlow (1982) predicts that massive tidal waves would sweep around the Earth when it tumbled, causing widespread destruction. (For a discussion of Warlow's hypothesis, see Slabinski 1981, 1982; and Warlow 1987.)

Global floods and superfast tumbles

A theme running through hypotheses of superfast tumble is that of sudden and violent global flooding. Certainly, a gradual change of sea level resulting from a small and relatively slow tumble is as nothing compared with the effects produced by a large, superfast tumble or full reversal. When, and if, the Earth turned upside-down, then massive tidal waves would sweep round the globe in a north–south, south–north direction surging over vast areas of land in the process. These tidal waves would, according to Warlow's analysis, be most destructive along the great circle that follows longitude 60 °W, 120 °E, because when the Earth reverses it would tend to do so, in the manner of a toy tippe-top, along a preferred axis. This axis is equatorial and passes through the middle of the Pacific Ocean and through the middle of Africa. Places on or near this axis would rotate in a relatively small arc and so the movement of waters would be relatively small. They would be 'safe areas'. Places which lie on or near the meridian of pole shift would bear the brunt of the tidal waves, and would suffer the worst destruction. It is interesting that the preferred meridian of pole shift during a tippe-top inversion, as defined by Warlow, is also the meridian of pole shift postulated by Weyer (1978) in his study of sea-level changes during the last Ice Age. However, the maximum rate of pole shift 'measured' by Weyer is roughly 0.5° a millenium, which is very much slower than the 180° a day predicted by Warlow.

The putative tidal waves envisaged by Velikovsky, Brown, and Warlow would have truly enormous erosive power. Warlow does not actually speculate on the effects of tidal waves on landforms, save to mention that they would carry vast quantities of sediment from the ocean floors and dump them on the continents (Warlow 1978: 2115). Brown and Velikovsky are more forthcoming on this matter. Velikovsky (1955), like Warlow, argues that the oceans would rush over the continents, depositing gravel, sand, and marine animals. He also suggests that lakes would be tilted and emptied, and that rivers would change their courses. Brown (1967) boldly asserts that vast areas of land would be gouged out by the rushing waters, leaving behind buttes, bluffs, and table mountains. He also offers what seems to be a novel, if unlikely, explanation of an erosion surface:

There is geological evidence that mountains have been cut off and carried away during the cataclysms of the deluges. Miles of . . . slanting rocks exist in normal formations which appear to have been clearly cut off. A fairly level plain is now all that remains where once a mountain stood at Joggins, Nova Scotia, on the Bay of Fundy . . . (Brown 1967: 50)

It is interesting that the scale of the cataclysms envisaged by the Velikovskians, with at the very least large areas of continental margins being flooded, could also be produced by the impact of a suitably large asteroid or comet in the ocean. The effects in both cases would be similar.

Fairbridge (1984) has ruled out Velikovskian models of Earth history because there is no evidence that the biosphere has ever been subjected to the devastation that superfast tumbles of the Earth would bring about. The logic behind this view seems sound: since it first evolved, the biosphere has always existed; therefore, it has never been utterly destroyed. Asimov (1979: 164) makes the same point when he asserts that there has been no catastrophe in the last four billion years which has been drastic enough to interfere with the development of life. None the less, Asimov and Fairbridge may be underestimating the resilience of the biosphere which, if the simulation studies made by Raup (1982) are a good guide, is a very robust beast. Raup simulated the effect of impacts on the geographical distribution of all families of living terrestrial vertebrates and all genera of two groups of marine organisms— corals and echinoids. Target points were selected at random on the Earth's surface and several lethal radii were specified. For each target point and for each lethal radius, a census was taken of those taxa which, because they happened to live in the lethal area, would become extinct. A lethal radius of 10 000 km covers roughly one hemisphere; a lethal radius of 20 000 km would encompass the entire Earth and lead to 100 per cent extinction. In terrestrial families, a total of 105 randomly chosen impact points were run, 30 times each, with lethal radii of 3000, 6000, and 15 000 km. For marine genera, 70 comparable simulations were performed. The results are shown in Figure 9.3. An important result of Raup's simulations is that the extinction rates are relatively low where lethal areas less than about half the Earth's surface area are involved. Destruction of all life over one full hemisphere (10 000 km lethal radius) eliminates on average 12 per cent of terrestrial families and 13.8 per cent of marine genera. These levels are below extinction levels typical of mass extinctions during the Phanerozoic. The conclusion is, therefore, at least using modern biogeographical patterns as the basis of the simulations, that mass extinctions require a global or near global crisis or deterioration of the environment.

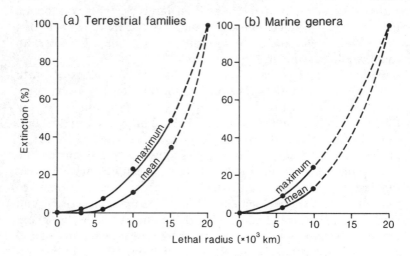

FIG. 9.3. Results of simulated biogeographical extinctions based on the present distribution of taxa. A lethal radius of 10 000 km is roughly equivalent to one-half of the Earth's surface area. A lethal radius of 20 000 km circumscribes the entire globe and would lead to 100 per cent extinction. (From Raup 1982)

The question that Raup was trying to solve was whether a purely regional biotic crisis produces enough extinctions for the event to qualify as mass extinction in a geological context. The answer seems to be no: a mass extinction requires either global environmental stress or a global catastrophe. Survival of species through a mass extinction may be the result of one or a few regions lucky enough to be unaffected by a near global catastrophe. This is just what would happen in the rapid pole shift scenarios, where 'safe areas' exist well away from the meridian of displacement. It would also happen in the superflood scenario, which will be described in a later section, where cataclysms are confined to the lands adjacent to an ocean, and do not cover the entire globe.

The bombardment hypothesis

Comets and catastrophes: early views

The notion that God might have used a comet as an instrument to flood the sinful, antediluvian world dates back many centuries. William Whiston, as was noted in Chapter 4, believed that the tail of a comet, coming into contact with the Earth in the year 2349 BC, had led to widespread flooding and wholesale extinction of animals, plants, and man. Edmund Halley, in a paper read to the Royal Society in 1694 and entitled 'Some considerations about the cause of the universal Deluge', proposed that a collision between the Earth and comet had been God's means of unleashing a cataclysm as enormous and powerful as Noah's (Halley 1724–5). And, at the conclusion of his classic paper on comets, Halley (1705) noted that the comet of 1680 had come close to the Earth and was prompted to write: 'But what might be the consequences of so near an appulse; or of a contact; or lastly, a shock of the celestial bodies, (which is by no means impossible to come to pass) I leave to be discussed by the studious of physical matters.' In 1755, Thomas Wright of Durham noted that it was 'not at all to be doubted from their vast magnitude and firey substance, that comets are capable of distroying such worlds as may chance to fall in their way' (quoted in Clube and Napier 1986: 261). This view was expanded upon by Pierre Simon, Marquis de Laplace (1749–1827) who, in his *Exposition du système du Monde* of 1796, claims that a comet encountering the Earth would cause cataclysmic events to occur. He writes of a change in the rotation axis and the direction of rotation which impart violent tremors to the globe, and cause the seas to abandon their basins and to precipitate themselves towards the new equator. He envisions a universal flood and massive earthquakes in which a great proportion of men and animals drown, entire species are wiped out, and all the monuments of human endeavour are destroyed. But these catastrophic prognostications were not widely accepted by the scientific intelligentsia of the Enlightenment, many of whom regarded the notion of celestial missiles as agents of catastrophism as a drawing-room joke (Clube and Napier 1986: 261). Cosmic catastrophism thus became regarded as improbable, a view which has persisted, and indeed been reinforced, during this century (Bailey *et al.* 1986: 91).

The impact crater controversy

The bombardment hypothesis started to be taken seriously when the first asteroid with an Earth-crossing orbit, 887 Alinda, was discovered by M. Wolf at Heidelberg in 1918 (Shoemaker 1983: 462). This asteroid had a diameter of about 5 km. In 1932, a second asteroid, 1221 Amor with a diameter of about 1 km, was discovered. The orbit of this asteroid was not thought at the time to cross the Earth's orbit, though it in fact does so. Also in 1932, the Earth-crossing asteroid 1862 Apollo was discovered. In the light of these and later discoveries of Earth-crossing asteroids, it became acceptable to suggest that stray meteorites might collide with the Earth. In 1942, H. H. Nininger of the Colorado Museum of Natural History and the American Meteorite Laboratory speculated on what would have happened had the asteroid Hermes, instead of passing by the Earth, as it had just done, had hit the Earth. He argued that a large meteorite impact would cause great changes in shorelines, the elevation and depression of extensive areas, the submergence of some low-lying areas of land, the creation of islands, withdrawal and extension of seas, and widespread and protracted volcanism; and that the collision between the Earth and planetoids offers an adequate explanation for the successive revolutions of movements in the Earth's crust which have been widely recognized, and for the sudden extinction of biota over large areas as revealed by the fossil record. Such suggestions as this were not taken seriously, probably because, interesting though they were as speculations, there seemed to be little evidence that meteorites had actually struck the Earth. True, a large and growing number of impact structures had been discovered, but their impact origin still remained not absolutely certain.

It was not until the early 1960s that Eugene M. Shoemaker and his colleagues developed a model of, and found unique evidence for, the impact origin of Meteor Crater, Arizona (Chao *et al*. 1960; Shoemaker 1963), and more or less settled a controversy which had raged for many decades (Hoyt 1987). Shoemaker had made detailed maps and structural analyses of Meteor Crater during the 1950s. In 1960, he sent a rock sample to E. C. T. Chao of the United States Geological Survey Laboratory in Washington, DC. The mineral coesite was detected in the sample. Further samples also proved to contain coesite. The discovery of coesite, a very dense and heavy form of silica which had been made in a laboratory under extremely high pressures by Loring Coes (1953), but was not known to occur in nature, was exciting and startling. Here in Meteor Crater was firm evidence supporting the view that the crater had been produced by a meteorite impact. Only a meteorite impact could produce high enough pressures for coesite to form. Shoemaker's discovery led to a search for coesite in other craters suspected of having an impact origin. The search was successful: coesite was found in rocks of the Ries Crater, West Germany, and at many other sites.

The proven association of coesite with impact-shocked rocks lent support for the view, first mooted by H. H. Nininger in 1956 and developed by M. E. Lipschutz and E. Anders (1961), that diamonds found in iron in the Canyon Diablo Crater were formed by an impact event. It became clear during the early 1960s that the alteration of minerals in target rock, induced by the passage of a shock wave radiating from the point of impact, was a sure signature of an impact event. The enormous pressures generated by a shock wave caused minerals to change instantaneously into glass without melting. Numerous examples of impact metamorphism, as it is known, have since been unearthed, and impact metamorphism is now taken as proof that a crater was produced by a meteorite impact.

An independent means of detecting and confirming the origin of impact craters was established by Robert S. Dietz. In 1947, Dietz published a paper in which he showed that shatter cones—conical fragments of rock with striations that radiate from the apex—were created by the impaction of meteorites at hypervelocities. Whether other geological processes could produce shatter cones was unclear. Certainly, shatter cones were not found in 'normal' rock formations, nor in rocks which had been subjected to volcanic explosions. Crude, irregular fracture cones without striations were produced by the explosives used in quarrying, while cones with striations, similar to shatter cones but less perfect in shape, were produced by military explosives with a high detonation velocity and high shattering effect. By the early 1960s, shatter cones at several impact sites had been discovered. They strongly suggested the occurrence of impacts, but did not provide unequivocal evidence since their origin was not fully understood, and it could not be shown that they could not be formed by other geological processes.

Thus it was that, from the foundations laid by Dietz, Shoemaker, and other pioneers, a rash of impact studies arose which eventually led to the general acceptance of the impact origin of the majority of lunar craters and their terrestrial counterparts. A few voices of dissent are sometimes heard (for example, Bucher 1963; McCall 1979), mainly because in larger craters no fragments of the impacting body remain, having been vaporized and melted on impact, but also because of the complexity of crater form at larger diameters. The dissenters have suggested a range of internal geological processes to account for crater formation: W. H. Bucher (1963), for instance, suggests crypto-explosions of gas. However, in the face of a voluminous literature on impact phenomena, all but a few geologists question the existence of large terrestrial impact structures lacking associated meteorite fragments (Grieve 1987; Mark 1987). Thus, owing to the convincing field evidence for impact events, and the other developments in space science mentioned in Chapter 1, the hypothesis of terrestrial catastrophism is now taken seriously by a large number of Earth scientists.

The effects of bombardment

The effects of bombardment are multifarious. Some are far better understood than others. One immediate consequence of an impact in the ocean is very clear, however. It is that enormous waves— superwaves—will be produced which will run out radially from the water cavity formed at the point of impact. On approaching continental margins, superwaves may grow to heights large enough to spill on to the land, flooding extensive areas of continental lowlands. The waters from these superfloods, in running back to the sea, would be capable of carrying out enormous amounts of work, possibly diverting rivers, cutting gorges, forming valleys, meanders, and perhaps even 'valleys of denudation', and laying down widespread sheets of gravel, the gravel having been produced by the initial force of an incoming superwave crashing against weathered bedrock (Huggett 1988, 1989). It is this kind of cataclysm which forms the basis of neodiluvialism, and it will now be explored in detail.

Superwaves and superfloods

Bombardment does occur. It will probably lead to superflooding on a continental, or possibly hemispherical, scale under two circumstances. Firstly, the impactor must hit the ocean. The chances of doing so are roughly 0.71 per impact. Secondly, the impactor must be large enough to survive a journey at hypervelocity through the atmosphere and hit the ocean with enough kinetic energy to generate a system of superwaves. As will be explained later, to produce a system of superwaves that would cause extensive superflooding, an impactor of at least ~0.5 km diameter is required. The present strike rate of impactors of that diameter or greater in the oceans is roughly between 10 and 18 strikes per million years. Clearly, therefore, superwaves should be generated fairly regularly over geological timespans.

Superwaves

To appreciate how superwaves are propagated, it is necessary to understand what happens to the ocean when it is struck at hypervelocity by an asteroid or comet. Any extraterrestrial body large enough to reach the Earth's surface will produce a crater. To survive the short journey through the atmosphere, an impactor must be at least 150 m in diameter (200 megatons), unless it be made of very dense iron when a smaller diameter will suffice. On hitting the Earth's surface, an impactor sets in train a sequence of events which involve the vaporization and ejection of surface material (rock or water) to leave a crater.

The sequence of events produced by an oceanic impact are clearly explained and vividly described by Emiliani *et al.* (1981), who consider the effects of an impactor of the same size as that which appears to have smitten the Earth 65 million years ago. The impactor has a mass of 2.5×10^{15} kg, a diameter of 14 km, and an impact velocity of 20 km/sec. The kinetic energy is thus approximately 5×10^{23} J. When struck by the impactor, the 5 km or thereabouts of ocean water acts like a thin, incompressible skin which is punctured. The water in the vicinity of the impact is highly compressed by the shock. It vaporizes upon decompression, spraying out of an expanding transient water cavity. Further away, weakly shocked water is also sprayed upwards. Additionally, a large volume of water is pushed out radially from the point of impact. The diameter of the transient cavity is about 140 km, although the sea water may well be swept away from the point of impact to give a water crater of around 200 km in diameter. The impact would breach the oceanic crust and upper mantle interface leaving a pronounced morphological, gravity, and magnetic structure—a hydrobleme. One such impact structure in the ocean floor has been discovered by Jansa and Pe-Piper (1987). And Kyte *et al.* (1988) have found evidence of the impact of a meteorite, 500 m in diameter, 2.3 million years ago in the south-east Pacific Ocean.

The shock of an impact in the ocean would, under some conditions, produce multiple wave systems. The initial wave system would be related to the transient rim of the water crater. The waves in such systems may have initial amplitudes as large as 4 to 5 km: they are indeed superwaves, a term coined by Gault and Sonett (1982).

Superwaves in scale experiments

The nature of the system of superwaves produced by oceanic impacts has been studied in laboratory experiments by Gault and Sonett (1982). The experiments, performed at the NASA Ames Research Center's Vertical Gun Ballistic Range, involved shooting spherical projectiles at velocities ranging from 1.25 to 6 km/sec into a tank of water. Oceanic impacts were thus scaled down through twenty orders of magnitude. Two sets of experiments were carried out: one set in 'deep' water, one set in 'shallow' water. The terms 'deep' and 'shallow' were defined in relation to the water depth, h, and the maximum depth of the transient cavity produced by the projectile, d_m. The ratio between the water depth and the maximum depth of the water cavity, β, was given by

$$\beta = h/d_m. \tag{1}$$

Gault and Sonett defined 'deep' as a β-value greater than 4. Under 'deep' conditions, the maximum depth of a cavity is attained without any damping or hindrance owing to the proximity of the ocean bottom. They define 'shallow' as a β-value in the range 0 to 1.5.

The 'deep'-water experiments established the following relationships between the maximum depth of the transient water cavity, d_m, the maximum diameter of the transient water cavity, D_m, and the kinetic energy of the projectile, w:

$$d_m = 0.1 \, w^{1/4} \tag{2}$$

$$D_m = 0.3 \, w^{1/4}. \tag{3}$$

where d_m and D_m are measured in cm, and w is measured in ergs. These relationships may be used to extrapolate the effects of large-body impacts. However, an S-type asteroid with a diameter of 0.2 km, impacting in the open ocean, will produce a water cavity about 2.64 km deep. This gives a β-value of $5/2.64 = 1.9$. Thus even a relatively small, large-body impact does not occur under 'deep'-water conditions where $\beta > 4$. Indeed, all oceanic impacts of asteroids, where the impactor has a diameter of more than 0.27 km, will give a β-value of less than 1.5, and so comply with the conditions defined as 'shallow' water by Gault and Sonett. The extrapolation of equations (2) and (3) to 'shallow' water conditions is possible, but it should be borne in mind that, if the diameter of the impactor is more than about 0.5 km, the maximum water crater depth cannot be realized, owing to the presence of the ocean floor, and the formation of waves will be affected. Gault and Sonett point out that all extrapolations may be invalid if, on impact, a fanlike jet of water or high-energy steam spreads nearly horizontally about the surface. This warning notwithstanding, an impact can be expected to produce a system of superwaves (Figure 9.4). An initial wave will form at the upraised rim of the water

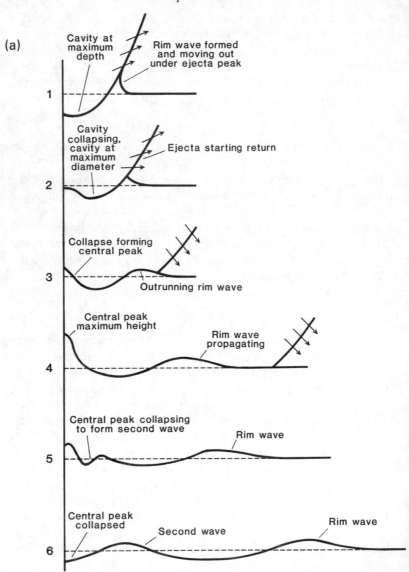

FIG. 9.4. The formation of superwaves in scale experiments. (a) Wave formation produced by a pyrex sphere with a diameter of 0.317 cm impacting in water at 2.31 km/sec. The β-value is 0.68. The time elapsed from the start to the end of the sequence is 0.65 seconds. (b) Wave

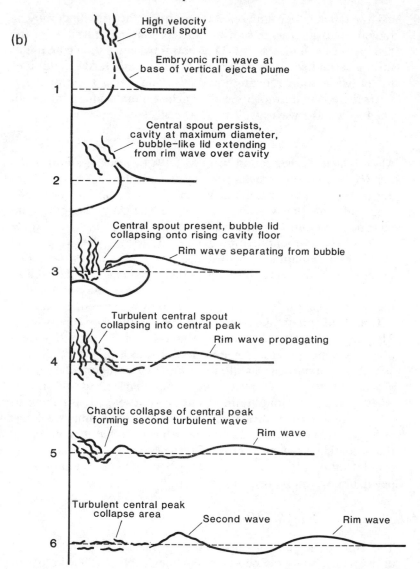

formation produced by an aluminium sphere with a diameter of 0.317 cm impacting in water at 5.64 km/sec. The β-value is 0.43. The time elapsed from the start to end of the sequence is 1.0 second. (After Gault and Sonett 1982)

cavity, quickly becoming detached from the rim and moving radially outwards as a central peak of water is formed by the collapse of the initial cavity. The central peak will then collapse to form a second wave. Whether there will be any further additions to the wave train remains unclear, though, if a hydrobleme is formed, the collapse of its walls might generate more waves. In deep water, the waves will move at velocity, v_w, defined as

$$v_w = \sqrt{(gh)}, \tag{4}$$

where g is the acceleration of gravity (980 cm/sec^2), and h is the depth of water. If $h = 5$ km (or 500 000 cm), then $v_w = 21\,908$ cm/sec or 0.22 km/sec maximum velocity. In moving outwards, the amplitude of the waves may be reduced by damping and dispersion, but the degree of reduction involved is unclear (Strelitz 1979). Even without damping and dispersion, the waves will diminish in amplitude with distance as $1/r$, where r is the radial distance from the wave source, and their wavelength will increase, because the mass of water in the wave system must be conserved as the wave radiation field spreads over an ever increasing area.

Gault and Sonett take a rather cautionary stance on superwaves. They emphasize that because speculations about amplitude and other parameters are so uncertain, little can be said with confidence about the shoaling of superwaves upon continental shelves. They believe that the key, as yet unanswered, question is whether the wave amplitude would be sufficient for breaking to take place far from the shore, where the continental shelf has already raised the ocean bottom to such a level that it is sensed by the oncoming wave train. However, they do admit that the energy content of superwaves, in cases where deep-sea damping is negligible, would be truly spectacular.

Superwave run-up height

The run-up heights of superwaves produced by impactors of different types and diameters is shown in Table 9.1. Two types of impactors were considered in constructing this table: low-density 'asteroids', and C-type and S-type asteroids (Table 9.2). The run-up heights were computed using the relationships expressed in equations (2) and (3), together with expressions for calculating asteroidal mass and the kinetic energy of impact (see Huggett

TABLE 9.1. *Superwave run-up height and impact size*

Impact energy (MT)	Scaled crater diameter (km)	Diameter of impactor			Superwave run-up height[a]	
		Low-density 'asteroid' (km)	C-type asteroid (km)	S-type asteroid (km)	1000 km from wave source (m)	3000 km from wave source (m)
457	4	0.33	0.22	0.19	10–100	3–34
71 400	9	0.83	0.55	0.49	39–390	13–130
103 000	17	1.71	1.34	1.01	67–670	22–220
234 000	25	2.65	1.76	1.57	93–930	31–310
7 760 000	70	8.50	5.65	5.04	220–2200	74–740

[a] All values assume that a superwave will grow in height by factor of 10 on approaching a coastline. The larger figure assumes that wave height decreases with distance from the point of impact purely because water volume is conserved in a spreading wave system. The lower figure, which is an order of magnitude lower than the higher one, allows for the additional reduction of wave height owing to damping and dispersion. In fact, the extent to which damping and dispersion will affect superwaves is very uncertain.

Source: 'Superwaves and superfloods: The bombardment hypothesis and geomorphology', *Earth Surface Processes and Landforms*, R. J. Huggett, Copyright 1989. Reprinted by permission of John Wiley & Sons, Ltd.

1989). Initial superwave amplitudes were set equal to the maximum depth of the water cavity, or depth to the ocean floor, whichever is the smaller, and using an inverse relationship between wave amplitude and radial distance from the point of impact. In order to allow for some reduction of superwave amplitude owing to turbulence and breaking in the vicinity of the wave source, and some damping and dispersion outside the immediate source area, two superwave amplitudes were calculated for a given radial distance from the point of impact: a 'maximum' amplitude, in which damping and dispersion were assumed to have no effect; and a 'minimum' amplitude, in which damping and dispersion were assumed to reduce the 'maximum' amplitude by an order of magnitude. The amplitude of actual superwaves was deemed to lie within the range between the 'maximum' and 'minimum'. The run-up heights of the superwaves were computed by supposing, with Gault *et al.* (1979), that wave height increases by a factor of ten on approaching shallow water.

At best, the results give a rough-and-ready guide to superwave run-up heights. None the less, as discussed in the next section, they do provide a useful starting-point from which a more detailed discussion of the significance of superwaves to geomorphology can develop.

Superfloods

If the foregoing analysis provides superwave run-up heights of the right order of magnitude, then the case for suggesting that superwaves will produce deep and widespread floods—super-floods—becomes clear. Other scientists have hinted that impact-induced waves will have catastrophic effects on plants, animals, and landforms. The astronomer H. H. Nininger, as long ago as 1942, commented that

If, instead of missing our planet by a few hundred thousand miles, the visit of the little planetoid Hermes had been timed slightly differently, a few billion tons of meteoritic material might have smacked the Earth in a single lump! Mathematicians tell us that a mass of iron, traveling at a speed of only 5 miles per second, would, upon its encounter with the lithosphere, become an explosive having about 4 times the violence of the same tonnage of nitroglycerin! . . . If an ocean were to receive such an explosive charge . . . the resultant tidal waves might prove quite as disastrous to land life on the adjoining continents as if the impact had occurred there. (Nininger 1942: 270)

TABLE 9.2. *Some properties of asteroids*

Type of body	Density, δ (g/cm^3)	Mean diameter, d (km)	Mean impact velocity, v (km/sec)	Population size,[a] N
Low-density 'asteroid'[b]	0.5	1.31	20.1	1300–2300
C-type asteroid[c]	1.7	1.73	20.1	650–1250
S-type asteroid[c]	2.4	0.89	20.1	650–1250

[a] The figures given are the minimum and maximum current estimates of the Earth-crossing asteroid population (Shoemaker 1983, 1984).

[b] Recent studies suggest that some 'asteroids' may be the disintegration products of giant comets and would therefore have densities in the range 0.1 to 1.0 g/cm^3 (Clube and Napier 1984; Napier, personal communication); a density of 0.5 g/cm^3 is assumed in this study.

[c] Data from Shoemaker (1983, 1984).

Source: 'Superwaves and superfloods: The bombardment hypothesis and geomorphology', *Earth Surface Processes and Landforms*, R. J. Huggett, Copyright 1989. Reprinted by permission of John Wiley & Sons, Ltd.

And Clube and Napier (1982: 103) point out that as superwaves approached land, a hydraulic bore of immense dimensions would be created leading to deep and catastrophic inundation. Similarly, Emiliani *et al.* (1981) explain that:

monstrous gravity waves would still have heights of several hundred meters, even on deep water many thousands of kilometers away. If the impact occurred in the Pacific these super tsunami would penetrate deeply into the surrounding continents with extremely destructive effects. This catastrophe by itself may have been sufficient to exterminate species whose habitat was restricted to the low lands around the northern Pacific. (Emiliani *et al.* 1981: 326)

An analysis of the expected strike rates by impactors of various sizes shows that, although the grand superfloods described by Nininger and others are fairly rare events, more moderate superfloods are, in geological time, common events, and strongly suggests that superfloods resulting from ocean impacts are of considerable importance in correctly interpreting some aspects of landscape history (Huggett 1989). The impaction rates of asteroids in the oceans are listed in Table 9.3. It is evident from the table, and follows from the size and frequency distribution of the Earth-crossing asteroid population, that big superfloods will be much less common than small superfloods. For example, superfloods produced by the impaction of asteroids with diameters of 0.2 km or more will be ~200 times more frequent than superfloods produced by the impaction of asteroids with a diameter of 5 km. It is also worth pointing out that the kinetic energy of an average S-type asteroid with a diameter of 0.2 km is about 500 megatons, which is about five times greater than the energy released by a big earthquake and ten times greater than the energy released in a large volcanic explosion. Thus, even small-body impacts are far more energetic than the sudden and violent process originating inside the Earth.

The impaction rates of asteroids in individual oceans are surprisingly high. For example, in the Pacific ocean, there will be between 23 and 41 strikes per million years by C-type and S-type asteroids, each capable of producing a crater of at least 4 km diameter on land. This rate is equivalent to one impact every ~24 000 to ~43 000 years. There will be between 5 and 9 strikes per million years by C-type and S-type asteroids each capable of producing a crater 9 km in diameter on land. This rate is

TABLE 9.3. *The impact rate of low-density, and S-type and C-type, asteroids in the oceans*

Scaled crater diameter (km)	Low-density asteroids		Superwave run-up height (m)		S-type and C-type asteroids	
	Oceanic impacts (m.yr.$^{-1}$)	Time between oceanic impacts (yr.)	1000 km from wave source	3000 km from wave source	Oceanic impacts (m.yr.$^{-1}$)	Time between oceanic impacts (yr.)
World oceans:						
4	21.2–37.7	47 000–26 000	10–100	3–34	45.9–81.8	21 000–12 000
9	4.7–8.4	210 000–110 000	39–390	13–130	10.2–18.2	97 000–54 000
17	1.4–2.6	670 000–380 000	67–670	22–220	2.6–4.7	370 000–210 000
25	0.72–1.2	1 300 000–770 000	93–930	31–310	1.5–2.7	640 000–360,000
70	0.10–0.19	9 100 000–5 100 000	220–2200	74–740	0.23–0.41	4 200 000–2 300 000
Pacific Ocean:						
4	10.6–18.9	94 000–52 000	10–100	3–34	23.0–41.0	43 000–24 000
9	2.3–4.2	410 000–230 000	39–390	13–130	5.1–9.1	190 000–100 000
17	0.73–1.3	1 300 000–750 000	67–670	22–220	1.3–2.3	750 000–420 000
25	0.36–0.64	2 700 000–1 500 000	93–930	31–310	0.78–1.3	1 200 000–710 000
70	0.055–0.098	18 000 000–10 000 000	220–2200	74–740	0.11–0.21	8 400 000–4 700 000
Atlantic Ocean:						
4	5.5–9.8	180 000–100 000	10–100	3–34	11.9–21.2	83 000–46 000
9	1.2–2.2	800 000–450 000	39–390	13–130	2.6–4.7	370 000–210 000

17	0.38–0.68	2 600 000–1 400 000	67–670	22–220	0.68–1.2	1 400 000–810 000
25	0.18–0.33	5 300 000–2 900 000	93–930	31–310	0.40–0.72	2 400 000–1 300 000
70	0.028–0.050	35 000 000–19 000 000	220–2200	74–740	0.061–0.10	16 000 000–9 100 000
Indian Ocean:						
4	4.3–7.7	230 000–120 000	10–100	3–34	9.4–16.7	100 000–59 000
9	0.97–1.7	1 000 000–570 000	39–390	13–130	2.1–3.7	470 000–260 000
17	0.30–0.53	3 300 000–1 800 000	67–670	22–220	0.54–0.96	1 800 000–1 000 000
25	0.14–0.26	6 700 000–3 700 000	93–930	31–310	0.31–0.56	3 100 000–1 700 000
70	0.022–0.040	44 000 000–24 000 000	220–2200	74–740	0.048–0.085	20 000 000–11 000 000
Arctic Ocean:						
4	0.72–1.2	1 300 000–770 000	10–100	3–34	1.5–2.7	640 000–350 000
9	0.16–0.28	6 100 000–3 400 000	39–390	13–130	0.34–0.62	2 800 000–1 600 000
17	0.050–0.089	19 000 000–11 000 000	67–670	22–220	0.089–0.16	11 000 000–6 200 000
25	0.024–0.043	40 000 000–22 00 000	93–930	31–310	0.052–0.094	18 000 000–10 000 000
70	0.003–0.006	267 000 000–150 000 000	220–2200	74–740	0.008–0.014	124 000 000–70 000 000
North Atlantic and Arctic Ocean affecting the British Isles:						
4	3.1–5.5	320 000–180 000	10–100	3–34	6.7–12.0	140 000–83 000
9	0.69–1.2	1 400 000–800 000	39–390	13–130	1.5–2.6	660 000–372 000
17	0.21–0.38	4 600 000–2 500 000	67–670	22–220	0.38–0.69	2 500 000–1 400 000
25	0.10–0.19	9 300 000–5 200 000	93–930	31–310	0.22–0.40	4 300 000–2 400 000
70	0.016–0.028	61 000 000–34 000 000	220–2200	74–740	0.034–0.061	28 000 000–16 000 000

Source: 'Superwaves and superfloods: The bombardment hypothesis and geomorphology', *Earth Surface Processes and Landforms*, R. J. Huggett, Copyright 1989. Reprinted by permission of John Wiley & Sons, Ltd.

equivalent to one impact every ~100 000 to ~190 000 years. These events are fairly frequent in geological terms, and should certainly be considered when trying to unravel the landscape history of continental lowlands surrounding the Pacific ocean. It is worth noting that an impact in the Pacific would produce synchronous landscape changes around the Pacific margins, but would leave no trace around the margins of the other ocean basins. This possible source of non-synchroneity of events on opposite sides of a continent should, perhaps, be borne in mind when attempting to correlate sedimentary sequences within, and between, continents.

Neodiluvialism

Superfloods and landscape change

It is virtually certain that a system of superwaves would be propagated by an asteroid (or comet) impacting in an ocean. It is less certain, but likely, that, on approaching land, these superwaves would grow to great heights. On reaching the shoreline, they would spill over the land at several hundred kilometres per hour, and would travel deep in to continental lowlands. Given the potential size of these superwaves and the sheer amount of energy carried by them, it is difficult to see how they could do anything but cause major changes in lowland landscapes, and if the waves are high enough, highland landscapes too. It would appear, then, that superwaves and associated superfloods may be an important factor in deciphering the landscape history of areas around continental margins. Certainly, all parts of a continent which are adjacent to oceans may be subject to periodic superflooding. Thus, there seems to be a case for reviving diluvialism in the guise of neodiluvialism.

It may be helpful at this juncture to consider what might happen to superwaves when they reach a coastline. Take the case of the British Isles (Huggett 1989). A low-density 'asteroid' with a diameter of 0.83 km striking the North Atlantic Ocean 1000 km west of Europe would produce superwaves which, on approaching the British Isles, would probably grow to between 39 and 390 m. A hit in the mid-Atlantic, 3000 km west of Britain, would produce superwaves that would probably grow to between 13 and 130 m on approaching British coasts. Events of this magnitude would occur, on average, once every 0.8 to 1.4 million years in the case of low-density 'asteroids', but more frequently—once every 372 000 to 660 000 years—in the case of S-type and C-type asteroids (Table 9.3). On striking the British coastline at up to 800 km/hr, a superwave 50 m or more high would be capable of stripping the soil cover from large areas, and, in places, fragmenting weathered bedrock to produce an ill-sorted mixture of boulders and gravels. And if it be doubted that superwaves of 50 m or less would leave their mark on the landscape, then consider the known effects of a modern floodwave that was produced by a vast rock-avalanche, triggered by an earthquake, in an inlet of the Gulf of Alaska in 1938. The displaced water in the inlet generated a wave with a steep front that rose to a height of 30 m or more and reached a velocity of 210 km/hr. The wave destroyed forest along kilometres of the shore and in places the momentum of the surging water carried it up to 525 m, as indicated by the height to which trees had been stripped of their bark and bedrock had been stripped of its soil cover (Holmes 1965: 905).

Superwaves generated by larger, but nevertheless relatively small, impacts—say those producing scaled crater diameters of at least 25 km—could grow to heights of 100 m or more. Waves of that size would surge inland, sweeping over lowland areas, causing a rapid and widespread inundation. Some of the waters of the superflood would probably rush over the country and continue to run into the North Sea. However a large body of water would temporarily remain in the chief river basins. Once the full system of superwaves had passed, the flood waters would drain away along existing drainage lines and into the sea. The amount of work done as the superflood waters ran back into the sea would be enormous. There would almost certainly be enough power available to divert rivers, cut deep gorges, produce valley meanders, effect widespread and intense erosion, possibly gouging out 'valleys of denudation', and lead to the deposition of extensive sheets of gravel. The gravel might possibly have been formed by the enormously high pressures exerted as superwaves crashed into weathered bedrock. Superfloods of this magnitude can be expected to occur in Britain about once every 2.4 to 9.3 million years (Table 9.3). Thus, they might have affected British landscapes during the last ten million years.

Even bigger superfloods in the British Isles are possible, but they will occur very infrequently (Table 9.3). Impacts which are capable of forming a crater with a diameter of 70 km or more occur once every 16 to 61 million years. They would generate superwaves which could grow to several hundred metres high, causing erosion in highland areas as well as in lowlands.

The signature of superfloods

The calculations presented in this chapter, even though they are very crude, would suggest that Quaternary landscapes were fashioned, in part, by the action of waters in superfloods. Highly speculative and ultra-controversial as this suggestion is, it can at least be tested: if superfloods have occurred, then they must surely have left their signature in the landscape. It will no doubt have occurred to readers that many of the landscape changes which may possibly be attributed to the action of superfloods, are the same as some of the landscape changes attributed to the Noachian Deluge by the old school of diluvialists led, at its brief but glorious nadir in the 1820s, by William Buckland. Of course, a number of the landscape features thought by the old diluvialists to have resulted from the action of diluvial waters have since been attributed to the action of ice. Nevertheless, not all the features mentioned by the diluvialists can be accounted for by the glacial theory. Valleys of denudation and gravel deposits, for instance, could be explained by the action of superfloods. There is, indeed, evidence for a regional cataclysm in 2300 BC (e.g. Mandelkehr 1983, 1987), which may correspond to the flood described in the Bible. It would be interesting to re-examine the landscape features and deposits described by the diluvialists in the light of modern techniques of sedimentology (e.g. Moffat and Catt 1986; Bridgland 1986) and palaeoflood analysis (e.g. Baker 1983; Baker and Pickup 1987; Partridge and Baker 1987; Stedinger and Baker 1987).

The modern catastrophists have repeatedly described the cataclysms that might have befallen the planet. In the light of the bombardment hypothesis, the case for looking again at the possibility that the landscape has in part been fashioned by the action of diluvial waters is persuasive. The theoretical argument in favour of neodiluvialism is a strong one. Whether the testimony of the rocks will bear out the neodiluvial theory has yet to be seen.

The time is come to go into the field and see if massive floods have left their signature in the landscape.

Conclusion

The time from the inscribing of the Epic of Gilgamesh on clay tablets to the late 1980s is geologically insigificant—a mere 4000 years or thereabouts. But in that brief period, human societies have grown rapidly, blooming into rich and varied civilizations. Almost all societies, both 'primitive' and 'advanced', have recorded, either in oral tradition or in writing, stories which hark back to episodes in Earth's violent past, to times when floods and fire wreaked havoc, causing catastrophes on a grand scale. Were the early poets and philosophers purveyors of actual historical events, or were they the first writers of science fiction? The unravelling of the threads of diluvialist thought has shown that, from the Renaissance to the early nineteenth century, the ancient flood myths were certainly widely accepted as records of real events, rather than allegories or products of the psyche. To most scholars of the Renaissance, Restoration, and Enlightenment, the Noachian Flood was an actual event of signal importance. This view rapidly faded during the nineteenth century as fewer and fewer geologists cared to countenance sudden and violent events as important factors in understanding Earth's surface history.

After the early 1860s, the fluvialistic view, which had been toyed with repeatedly since Classical times, rapidly took over the minds of geologists and geomorphologists to the exclusion of all other systems of Earth history. During the heyday of fluvialism, anybody, like J. Harlen Bretz, rash enough to propose that certain features of the landscape might have been fashioned by sudden and violent floods, was castigated in an inexcusably high-handed manner. The fluvialists, though they did have some spectacular achievements, evidently suffered from a different species of mental derangement from that diagnosed by Playfair in the cosmogonists. In the fluvialists' case it was their vision that was impaired, for they could only see along a narrow tunnel, as well as their minds, for they kept blathering about the present being the key to the past. This is a caricature, of course; but, like all caricatures, it contains a grain of truth. An unproductive achievement of the fluvialists was to replace one dogma with another.

However, geological and geomorphological fashions, like events in Earth's history, have a habit of repeating themselves. Catastrophism started to make a comeback in the 1970s. Diluvialism has recently followed suit. Of course, neocatastrophism and neodiluvialism are very different beasts from their predecessors. Neodiluvialism is not based on the hearsay evidence of ancient chroniclers, nor on the seemingly erroneous evidence of 'glacial' drift and erratics. It is firmly based on the latest findings in theoretical astronomy and comparative planetology. And neodiluvialism does not stand in juxtaposition to fluvialism. Rather, it allows that gradual and gentle processes do modify the Earth's surface, but it also claims that some topographical features and deposits may be produced by cataclysms so grand that they dwarf the Spokane Flood.

By considering neodiluvialism and fluvialism together, a better understanding of the history of the Earth's surface may be forthcoming. To advocate dogmatically either neodiluvialism or fluvialism is foolish, for both are valid systems of Earth history, and both may be used to explain certain Earth surface phenomena. They are not mutually exclusive systems: one does not have to be right and the other wrong. This book has tried to show why diluvialism should not be dismissed as an historical curiosity, and how some of the ideas of the old diluvialists live on in neodiluvialism. It has shown how gradualistic diluvialism—the relatively gradual flooding of continents associated with marine transgressions—has been widely identified and fairly satisfactorily explained, though the effect of transgressions on landscape development is far from being fully understood and deserves further investigation. It is far too early to predict that neodiluvialism will provide new explanations of many Earth surface features and deposits. But if it should achieve nothing more than making the more blinkered fluvialists question some of their most cherished beliefs, it will have done a good service to the Earth surface sciences. If it should prove to be a valuable system of Earth surface history, then that will be a bonus. The fact is that, for better or worse, diluvialism is in the throes of rebirth. Whether it will mature and bear fruit is difficult to say. The signs so far are auspicious.

References

ADAMS, F. D. (1938), *The birth and development of the geological sciences*, Baillière, Tindall, and Cox, London.

ADHÉMAR, J. A. (1842), *Révolutions de la mer*, Carilan-Goeury et V. Dalmont, Paris.

AGASSIZ, L. (1840), *Études sur les glaciers*, privately published, Neuchâtel.

ALESSANDRI, A. DEGLI (1522), *Dies geniales* (see Lyell 1834, vol. i; Gortani 1963).

ANDERSON, I. (1986), 'A glimpse of the Green Hills of Antarctica', *New Scientist*, 111: 22.

ANDREWS, J. A. (1985), 'True polar wander: An analysis of Cenozoic and Mesozoic palcomagnetic poles', *Journal of Geophysical Research*, 90: 7737–50.

ANON. (1820), 'Reflections on the Noachian Deluge, and on the attempts lately made at Oxford, for connecting the same with present geological appearances', *Philosophical Magazine and Journal*, 56: 10–14.

ARISTOTELES (1930), 'Physica', in *The works of Aristotle,* vol. ii, trans. R. P. Hardie and R. K. Gaye, Clarendon Press, Oxford, pp. 184–267.

—— (1931), 'Meteorologica', in *The works of Aristotle*, vol. iii, trans. E. W. Webster, Clarendon Press, Oxford, pp. 338–90.

ASIMOV, I. (1979), *A choice of catastrophes*, Hutchinson, London.

BAILEY, M. E., CLUBE, S. V. M., and NAPIER, W. M. (1986), 'The origin of comets, *Vistas in Astronomy*, 29: 52–112.

BAKER, V. R. (1973), 'Paleohydrology and sedimentology of Lake Missoula flooding in eastern Washington', *Geological Society of America Special Paper* 144.

—— (1977), 'Stream channel response to floods with examples from central Texas', *Bulletin of the Geological Society of America*, 88: 1057–71.

—— (1978), 'The Spokane Flood controversy and Martian outflow channels', *Science,* 202: 1249–56.

—— (1983), 'Large-scale fluvial palaeohydrology', in K. J. Gregory (ed.), *Background to palaeohydrology*, Wiley, Chichester, pp. 453–78.

—— and PICKUP, G. (1987), 'Flood geomorphology of the Katherine Gorge, Northern Territory, Australia', *Bulletin of the Geological Society of America*, 98: 635–46.

BARLEY, A. H. (1922), *The Drayson problem: An astronomical survey of the whole question, in the form of a reply to a recent article in the Journal of the British Astronomical Association entitled 'The Draysonian fallacy'* [a companion to R. A. Marriott, *The Ice Age fully explained*], William Pollard, Exeter.

BARNES, J. (1987), *Early Greek philosophy*, Penguin Books, London.

BAULIG, H. (1935), 'The changing sea level', *Transactions of the Institute of British Geographers*, 3: 1–46.

BELT, T. (1874), 'An examination of the theories that have been proposed to account for the climate of the glacial period', *Quarterly Journal of Science*, 44: 6–44.

BENSON, R. H. (1984), 'Perfection continuity, and common sense in historical geology', in W. A. Berggren and J. A. Van Couvering (eds.), *Catastrophes and Earth history: The new uniformitarianism*, Princeton University Press, Princeton, NJ, pp. 35–75.

BLANCHARD, J. (1942), *L'Hypothèse du déplacement des pôles et la chronologie du Quaternaire*, C. Monnoyer, Le Mans.

BONNET, C. (1779–83), *Œuvres d'histoire naturelle et de philosophie de Charles Bonnet, . . .*, 18 vols., Samuel Fauché, Neuchâtel.

BOULANGER, N.-A. (1766), *L'Antiquité dévoilée par sea usages, ou examen critique des principales opinions, cérémonies et institutions religieuses et politiques des différens peuples de la terre*, Marc-Michel Rey, Amsterdam.

BOYD, H. S. (1817), 'On cosmogony', *Philosophical Magazine and Journal*, 50: 375–8.

BOYLE, R. (1772), *The works of the Honourable Robert Boyle, . . . To which is prefixed the life of the author*, by Thomas Birch, 6 vols., J. & F. Rivington, London.

BRETZ, J. H. (1923*a*), 'Glacial drainage on the Columbia Plateau', *Bulletin of the Geological Society of America*, 34: 573–608.

—— (1923*b*), 'The Channeled Scabland of the Columbia Plateau', *Journal of Geology*, 31: 617–49.

—— (1928), 'The Channeled Scabland of eastern Washington', *Geographical Review*, 18: 446–77.

—— (1978), 'Introduction', in V. R. Baker and D. Nummedal (eds.), *The Channeled Scabland*, National Aeronautics and Space Administration, Washington, DC, pp. 1–2.

—— SMITH, H. T. U., and NEFF, G. E. (1956), 'Channeled Scabland of Washington: New data and interpretations', *Bulletin of the Geological Society of America*, 67: 957–1049.

BRIDGLAND, D. R. (ed.) (1986), *Clast lithological analysis*, Quaternary Research Association Technical Guide No. 3.

BROCCHI, G. B. (1814), *Conchyliologia fossile subappenia*, 2 vols., Milan.

BROWN, E. H., and WATERS, R. S. (1974), 'Geomorphology in the United Kingdom since the First World War', in E. H. Brown and R. S. Waters (eds.), *Progress in geomorphology: Papers in honour of David L. Linton*, Institute of British Geographers Special Publication No. 7, Institute of British Geographers, London, pp. 3–9.

BROWN, H. A. (1948), *Popular awakening concerning the impending flood*, published by the author.

—— (1967), *Cataclysms of the Earth*, Twayne, New York.

BUCHER, W. H. (1963), 'Cryptoexplosion structures from without or within the Earth ("Astroblemes" or "Geoblemes?")', *American Journal of Science*, 261: 596–649.

BUCKLAND, W. (1820), *Vindiciae geologicae: Or the connexion of geology with religion explained, in an inaugural lecture delivered before the University of Oxford, May 15, 1819, on the endowment of a Readership in Geology by his Royal Highness the Prince Regent*, published for the author by the University Press, Oxford.

—— (1823), *Reliquiae diluvianae: Or, observations on the organic remains contained in caves, fissures, and diluvial gravel, and on other geological phenomena, attesting the action of an universal deluge*, John Murray, London.

—— (1824a), *Reliquiae diluvianae: Or, observations on the organic remains contained in caves, fissures, and diluvial gravel, and on other geological phenomena, attesting the action of an universal deluge*, 2nd edn., John Murray, London.

—— (1824b), 'On the excavation of valleys by diluvian action, as illustrated by a succession of valleys which intersect the south coast of Dorset and Devon', *Transactions of the Geological Society*, 2nd ser., 1: 95–102.

—— (1829), 'On the formation of the valley of Kingsclere and other valleys, by the elevation of the strata that enclose them; and on the evidences of the original continuity of the basins of London and Hampshire', *Transactions of the Geological Society*, 2nd ser., 2: 119–30.

—— (1836), *Geology and mineralogy considered with reference to natural theology*, Treatise VI of The Bridgewater treatises on the power, wisdom, and goodness of God as manifested in the Creation, 2 vols., William Pickering, London.

BUFFON, G. L. L. DE (1749–89), *Histoire naturelle, générale et particulière, avec la description du Cabinet du Roi* (by Buffon, Daubenton, Guéneau de Montbeillard, Bexan, and Lacépède), 44 vols., De l'Imprimerie Royale, Paris.

—— (1778), *Les époques de la nature*, De l'Imprimerie Royale, Paris.

BURNET, T. (1681), *Telluris theoria sacra, originem et mutationes generales orbis nostri, quas aut jam subiit, aut olim subiturus est, complectens:*

Accedunt archaeologicae philosophicae, sive doctrina antiqua de rerum originibus, 2 vols., Walter Kettilby, London.

—— (1691), *The sacred theory of the Earth: Containing an account of the original of the Earth, and of all the general changes which it hath already undergone, or is to undergo, till the consummation of all things*, 2nd edn., Walter Kettilby, London. [The two first books, concerning the Deluge, and Paradise were published in 1691; the two last books, concerning the burning of the world, and concerning the new heavens and new Earth, were published in 1690.]

—— (1965), *The sacred theory of the Earth*, with an introduction by Basil Willey, Centaur Press, London (repr. of 2nd edn.).

BURY, H. (1923–4), 'The Bournemouth plateau and its palaeoliths', *Proceedings of the Bournemouth Natural Science Society*, 16: 72–81.

CALVIN, J. (1948), *Commentaries on the First Book of Moses called Genesis*, trans. John King, Eerdmans, Grand Rapids, Mich.

CARDANO, GIROLAMO (1550), *De subtilitate libri xxi, ab authore plusquam mille locis illustrati, nonnullis etiam cum additionibus: Addita insuper apologia adversus calumniatorem, qua vis horum librorum aperitur*, ex officio Petrina, Basileae.

CARPENTER, N. (1625), *Geography delineated forth in two bookes: Containing the sphaericall and tropicall parts thereof*, Henry Cripps, Oxford.

CARR, J. (1809a), 'On the natural causes which operate in the formation of valleys', *Philosophical Magazine*, 33: 452–9.

—— (1809b), 'On the causes which have operated in the production of valleys', *Philosophical Magazine*, 34: 190–200.

CATCOTT, A. (1761), *A treatise on the Deluge, containing I. Remarks on the Lord Bishop Clogher's remarks on that event. II. A full explanation of the scriptural history of it. III. A collection of all the principal heathen accounts. IV. Natural proofs of the Deluge, deduced from a great variety of circumstances, on and in the terraqueous globe*, M. Withers, London.

CESALPINO, A. (1596), *De metallicis libri tres*, A. Zanetti, Rome.

CHADWICK, P. (1962), 'Mountain-building hypotheses', in S. K. Runcorn (ed.), *Continental drift*, Academic Press, New York and London, pp. 195–234.

CHAMBERS, R. (1848), *Ancient sea margins, as memorials of changes in the relative level of land and sea*, W. and R. Chambers, Edinburgh, W. S. Orr, London.

CHAO, E. C. T., SHOEMAKER, E. M., and MADSEN, B. M. (1960), 'First natural occurrence of coesite', *Science*, 132: 220–2.

CHORLEY, R. J. (1965), 'The application of quantitative methods to geomorphology', in R. J. Chorley and P. Haggett (eds.), *Frontiers in geographical teaching: The Madingley Lectures for 1963*, Methuen, London, pp. 147–63.

—— Dunn, A. J., and Beckinsale, R. P. (1964), *The history of the study of landforms of the development of geomorphology*, vol. i: *Geomorphology before Davis*, Methuen Wiley, London.

Clayton, R. (1752), *A vindication of the histories of the Old and New Testament in answer to the objections of the late Lord Bolingbroke: In two letters to a young nobleman*, George Faulkener, Dublin.

Clube, S. V. M., and Napier, W. M. (1982), *The cosmic serpent: A catastrophist view of Earth history*, Faber and Faber, London.

—— —— (1984), 'The microstructure of terrestrial catastrophism', *Monthly Notices of the Royal Astronomical Society*, 211: 953–968.

—— —— (1986), 'Giant comets and the Galaxy: implications of the terrestrial record', in R. Smoluchowski, J. N. Bahcall, and M. S. Matthews (eds.), *The Galaxy and the Solar System*, The University of Arizona Press, Tucson, Ariz., pp. 260–85.

Cockburn, P. (1750), *An enquiry into the truth and certainty of the Mosaic Deluge: Wherein the arguments of the learned Isaac Vossius, and others, for a topical Deluge are examined; and some vulgar errors, relating to that grand catastrophe, are discover'd*, C. Hitch, London, M. Bryson, Newcastle upon Tyne.

Coes, L. Jun. (1953), 'A new dense crystalline silica', *Science*, 118: 131–2.

Collier, K. B. (1934), *Cosmogonies of our fathers: Some theories of the seventeenth and eighteenth centuries*, Columbia University Press, New York.

Colonna, Fabio (1616), *Osservazioni sugli animali aquatici et terrestri* (see Lyell 1834, vol. i: 38–9; Gortani 1963).

Conybeare, W. D. (1830–1), 'An examination of those phaenomena of geology, which seem to bear most directly on theoretical speculations', *Philosophical Magazine and Annals of Philosophy*, ns, 8 (1830): 359–62, and 401–6; 9 (1831), 19–23, 111–17, 188–97, 258–70.

—— (1834), 'On the valley of the Thames', *Proceedings of the Geological Society of London*, 2nd ser., 1: 145–9.

Cotta, B. (1846), *Grundriss der Geognosie und Geologie*, Arnoldische Buchhandlung, Dresden and Leipzig.

Courtillot, V., and Besse, J. (1987), 'Magnetic field reversals, polar wander, and core-mantle coupling', *Science*, 237: 1140–7.

Croll, J. (1865), 'On the physical cause of the submergence of the land during the glacial epoch', *The Reader: A Review of Literature, Science, and Art*, 6: 270–1.

Cusanus, N. (1954), *Of learned ignorance*, trans. Germain Heron, Yale University Press, New Haven, Conn.

Cuvier, G. (1812a), *Recherches sur les ossemens fossile, où l'on rétablit les caratères de plusiers animaux dont les révolutions du globe ont détruit les espèces*, 4 vols., Deterville, Paris.

—— (1812b), *Discours sur les révolutions de la surface du globe, et sur les changements qu'elles ont produits dans le règne animal*, (*Discours preliminaire* of above), Deterville, Paris.

—— (1822), *Essay on the theory of the Earth: With mineralogical illustrations by Professor Jameson*, 4th edn., with additions, trans. Robert Kerr, William Blackwood, Edinburgh, and T. Cadell, London.

DANIEL, G. E. (1959), 'The idea of man's antiquity', *Scientific American*, 201: 167–76.

—— (1981), *A short history of archaeology*, Thames and Hudson, London.

DARWIN, C. R. (1839), 'Observations on the parallel roads of Glen Roy with an attempt to prove that they are of marine origin', *Philosophical Transactions of the Royal Society*, 129: 39–81.

—— (1859), *The origin of species by means of natural selection, or the preservation of favoured races in the struggle for life*, John Murray, London.

DARWIN, G. H. (1877), 'On the influence of geological changes on the Earth's axis of rotation', *Philosophical Transactions of the Royal Society*, 167: 271–312.

—— (1880), 'On the secular changes in the elements of the orbit of a satellite revolving about a tidally distorted planet', *Philosophical Transactions of the Royal Society*, 171: 713–891.

DAVIES, G. L. (1964), 'Robert Hooke and his conception of Earth-history', *Proceedings of the Geologists' Association, London*, 75: 493–8.

—— (1969), *The Earth in decay: A history of British geomorphology, 1578–1878*, MacDonald, London.

DA VINCI, LEONARDO (1977), *The notebooks of Leonardo da Vinci*, arranged, rendered into English, and introduced by Edward McCurdy, 2 vols., Jonathan Cape, London.

DE HORSEY, A. F. R. (1911), *Draysonia: Being an attempt to explain and popularise the system of the second rotation of the earth as discovered by the late Major-General A. W. Drayson F.R.A.S. for fifteen years Professor Royal Military Academy, Woolwich. Also giving the probable date and duration of the last glacial period, and furnishing General Drayson's data, from which any person of ordinary mathematical ability is enabled to calculate the obliquity of the ecliptic, the precession of the equinoxes, and the right ascension and declination of the fixed stars for any year, past, present, or future*, Longman, Green, London.

DE LA BECHE, H. T. (1834), *Researches in theoretical geology*, Charles Knight, London.

—— (1849), 'Anniversary address of the President', *The Quarterly Journal of the Geological Society of London*, 5: pp. xix–cxvi.

—— (1851), *The geological observer*, Longman, Brown, Green, and Longmans, London.

DE LUC, J. A. (1778), *Lettres physique et morales sur l'histoire de la terre et de l'homme: Adressées à M. le Professor Blumenbach, renfermant des nouvelles preuves géologiques et historiques de la mission divine de Moyse*, Nyon, Paris. [Also published in the *British Critic*, 1793–5.]

DENTON, G. H., and HUGHES, T. J. (1983), 'Milankovitch theory of ice ages: Hypotheses of ice-sheet linkage between regional insolation and global climate', *Quarternary Research*, 20: 125–44.

—— ——, and KARLÉN, W. (1986), 'Global ice-sheet system interlocked by sea level', *Quaternary Research*, 26: 3–26.

DESCARTES, R. (1644), *Principia philosophiae*, Apud Ludovicum Elzevirium, Amsterdam.

DESMAREST, N. (1794–1828), *Géographie physique, (Encyclopédie méthodique)*, 5 vols., Paris.

DIETZ, R. S. (1947), 'Meteorite impact suggested by orientation of shatter cones at the Kentland, Indiana, disturbance', *Science*, 105: 42–3.

DOLOMIEU, D. G. S. T. G. DE (1791), 'Mémoire sur les pierres composées et sur les roches', *Observations sur La Physique, sur L'Histoire Naturelle et sur Les Arts*, 39: 374–407.

DRAYSON, A. W. (1871), 'On the cause, date, and duration of the glacial epoch of geology', *The Quarterly Journal of the Geological Society of London*, 27: 232–4.

—— (1873), *On the cause, date, and duration of the last glacial epoch of geology, and the probable antiquity of man: With an investigation and description of a new movement of the Earth*, Chapman and Hall, London.

DURY, G. H. (1953), 'The shrinkage of the Warwickshire Itchen', *Proceedings of the Coventry Natural History and Science Society*, 2: 208–14.

—— (1964), 'Principles of underfit streams', *US Geological Survey Professional Paper* 425.

—— (1965), 'General theory of meandering valleys', *US Geological Survey Professional Paper* 452.

—— (1969), 'Relation of morphometry to runoff frequency', in R. J. Chorley, (ed.), *Water, Earth, and man: A synthesis of hydrology, geomorphology, and socio-economic geography*, Methuen, London, pp. 419–30.

EGYED, L. (1956a), 'Determination of changes in the dimensions of the Earth from palaeogeographical data', *Nature*, 178: 543.

—— (1956b), 'The change of the Earth's dimensions determined from palaeogeographical data', *Geofisica Pura e Applicata*, 33: 42–8.

ÉLIE DE BEAUMONT J. B. A. L. L. (1831) 'Researches on some revolutions which have taken place on the surface of the globe; presenting various examples of the coincidence between the elevation of beds in certain systems of mountains, and the sudden changes which have produced the lines of demarcation observable in certain stages of sedimentary deposits', *Philosophical Magazine*, NS, 10: 241–64.

—— (1852), *Notice sur les systèmes des montagnes*, 3 vols., L. Martinet, Paris.

EMILIANI, C., KRAUS, E. B., and SHOEMAKER, E. M. (1981), 'Sudden death at the end of the Mesozoic', *Earth and Planetary Science Letters*, 55: 317–34.

EVANS, J. (1866), 'On a possible geological cause of changes in the position of the axis of the Earth's crust', *Proceedings of the Royal Society*, 82: 46–54.

—— (1876), 'The anniversary address of the President', *The Quarterly Journal of the Geological Society of London*, 32: 53–121.

EVANS, L. (1755), *An analysis of a general map of the Middle British Colonies in America*, B. Franklin and D. Hall, Philadelphia.

EYLES, V. A. (1969), 'The extent of geological knowledge in the eighteenth century, and the methods by which it was diffused', in C. J. Schneer (ed.), *Toward a history of geology*, The MIT Press, Cambridge, Mass., pp. 159–83.

FAENZI, V. (1561), *De montium origine, Valerii Faventies, ordinis praedicatorum, dialogus*, Aldine Press, Venice.

FAIRBRIDGE, R. W. (1961), 'Eustatic changes in sea level', *Physics and Chemistry of the Earth*, 4: 99–185.

—— (1970a), 'South Pole reaches the Sahara', *Science*, 168: 878–81.

—— (1970b), 'An ice-age in the Sahara', *Geotimes*, 15: 18–20.

—— (1971), 'Upper Ordovician glaciation in northwest Africa? Reply', *Bulletin of the Geological Society of America*, 82: 269–74.

—— (1984), 'Planetary periodicities and terrestrial climate stress', in N.-A. Mörner and W. Karlén (eds.), *Climatic changes on a yearly to millenial basis*, Reidel, Dordrecht, pp. 509–20.

FAUL, H., and FAUL, C. (1983), *It began with a stone: A history of geology from the Stone Age to the Age of Plate Tectonics*, Wiley, New York.

FELLOWS, O. E., and MILLIKEN, S. F. (1972), *Buffon*, Twayne's World Authors Series (TWAS 243), Twayne, New York.

FISHER, O. (1878a), 'On the possibility of changes in the latitudes of places on the Earth's surface; being an appeal to physicists', *Geological Magazine*, Decade 2, 5: 291–7.

—— (1878b), 'On the possibility of changes in the latitudes of places on the Earth's surface; being a reply to Mr. Hill's letter', *Geological Magazine*, Decade 2, 5: 551–2.

FLEMING, J. (1826), 'The geological deluge, as interpreted by Baron Cuvier and Professor Buckland, inconsistent with the testimony of Moses and the phenomena of nature', *Edinburgh New Philosophical Journal*, 14: 205–39.

FORCE, J. E. (1983), 'Linking history and rational sciences in the

Enlightenment', introduction to reprint edn. of William Whiston's *Astronomical principles of religion, natural and reveal'd*, Georg Orms Verlag, Hildesheim, Zurich, and New York.

FRAPOLLI, L. (1846–7), 'Réflexions sur la nature et sur l'application du charactère géologique', *Bulletin de la Société Géologique de France*, 4: 623–5.

FREUND, P. (1964), *Myths of creation*, W. H. Allen, London.

FÜCHSEL, G. C. (1762), *Historia terrae et maris, ex historia Thuringiae, per montium descriptionem*, Akademie gemeinnütziger Wissenschaften zu Erfurt Acta, 2, 44–209.

—— (1773), *Entwurfe der ältesten Erd- und Menschengeschichte, nebst Versuch, den Ursprung der Sprache zu finden*, Frankfurt.

GASCAR, P. (1983), *Buffon*, Gallimard, l'Imprimerie Floch, Mayenne.

GAULT, D. E., SONETT, C. P., and WEDEKIND, J. A. (1979), 'Tsunami generation by pelagic planetoid impact (abstract);', *Lunar Planetary Science Conference, X*, Lunar Planetary Science Institute, Houston, pp. 422–4.

—— —— (1982), 'Laboratory simulation of pelagic asteroidal impact: atmospheric injection, benthic topography, and the surface wave radiation field', in L. T. Silver and P. II. Schultz (eds.), *Geological implications of impacts of large asteroids and comets on the Earth*, Geological Society of America Special Paper 190, pp. 69–92.

GEIKIE, A. (1905), *The founders of geology*, 2nd edn., Macmillan, London.

GEORGE, T. N. (1974), 'Prologue to a geomorphology of Britain', in E. H. Brown and R. S. Waters (eds.), *Progress in geomorphology: Papers in honour of David L. Linton*, Institute of British Geographers Special Publication No. 7, Institute of British Geographers, London, pp. 113–25.

GESNER, ABRAHAM (1836), *Remarks on the geology and mineralogy of Nova Scotia*, Gossip and Coade, Halifax, Nova Scotia.

GIBSON, J. B. (1836), 'Remarks on the geology of the lakes and the valley of the Mississippi suggested by an excursion to the Niagara and Detroit Rivers, in July, 1833', *American Journal of Science*, 29: 201–13.

GOLD, T. (1955), 'Instability of the Earth's axis of rotation', *Nature*, 175: 526–9.

GOLDREICH, P., and TOOMRE, A. (1969), 'Some remarks on polar wandering', *Journal of Geophysical Research*, 74: 2555–67.

GORTANI, M. (1963), 'Italian pioneers in geology and mineralogy', *Cahiers d'Histoire Mondial*, 7: 503–19.

GOULD, S. J. (1965), 'Is uniformitarianism necessary?', *American Journal of Science*, 263: 223–8.

—— (1987), *Time's arrow, time's cycle: Myth and metaphor in the discovery of geological time*, Harvard University Press, Cambridge, Mass., and London.

GREENOUGH, G. B. (1819), *A critical examination of the first principles of geology; in a series of essays*, Longman, Hurst, Rees, Orme, and Brown, London.

—— (1833–4), 'Address delivered at the Anniversary Meeting of the Geological Society, on the 21st of February 1834, by George Bellas Greenough, Esq. President', *Proceedings of the Geological Society of London*, 2: 42–70.

GREENWOOD, G. (1877), *River terraces: Letters on geological and other subjects*, ed. Charles W. Greenwood with a memoir of the author by George Greenwood, Longmans, Green, London.

GRIEVE, R. A. F. (1987), 'Terrestrial impact structure', *Annual Reviews of Earth and Planetary Sciences*, 15: 245–70.

HALL, BASIL (1841), *Patchwork*, 3 vols., E. Moxon, London.

HALL, J. (1812), 'On the revolutions of the Earth's surface', *Transactions of the Royal Society of Edinburgh*, 7: 139–212.

HALLAM, A. (1963), 'Major epeirogenic and eustatic changes since the Cretaceous, and their possible relationship to crustal structure', *American Journal of Science*, 261: 397–423.

—— (1971), 'Re-evaluation of the palaeogeographical argument for an expanding Earth', *Nature*, 232: 180–2.

—— (1977), 'Secular changes in marine inundation of the USSR and North America during the Phanerozoic', *Nature*, 269: 769–72.

—— (1978), 'Seismic stratigraphy and global changes', *Nature*, 272: 400.

—— (1983), *Great geological controversies*, Oxford University Press, Oxford.

HALLEY, E. (1705), *A synopsis of the astronomy of comets*, John Senex, London.

—— (1724–5), 'Some considerations about the cause of the universal Deluge', *Philosophical Transactions*, 33: 118–25.

HAPGOOD, C. H. (1958), *Earth's shifting crust: A key to some basic problems of Earth science*, Pantheon, New York.

—— (1970), *The path of the pole*, Chilton, Philadelphia.

HARGREAVES, R. B., and DUNCAN, R. A. (1973), 'Does the mantle roll?', *Nature*, 245: 361–3.

HAYDEN, H. H. (1820), *Geological essays*, Robinson, Baltimore, Md.

HAYES, G. E. (1839), 'Remarks on the geology and topography of western New York', *American Journal of Science*, 35: 86–105.

HAYS, J. D., and PITMAN III, W. C. (1973), 'Lithospheric plate motion, sea level changes and climatic and ecological consequences', *Nature*, 246: 18–22.

HENSLOW, J. S. (1823), 'On the Deluge', *The Annals of Philosophy*, NS, 6: 344–8.

HILL, E. (1878*a*), 'On the possibility of changes in the Earth's axis', *Geological Magazine*, Decade 2, 5: 262–6.

—— (1878*b*), 'The possibility of changes of latitude', *Geological Magazine*, Decade 2, 5: 479.

HIPPOLYTUS (1870), *The refutation of all heresies by Hippolytus*, trans. J. H. MacMahon, 2 vols., T. & T. Clark, Edinburgh.

HITCHCOCK, E. (1819), 'Remarks on the geology and mineralogy of a section of the Massachusetts on the Connecticut River, with a part of New-Hampshire and Vermont', *American Journal of Science*, 1: 105–16.

—— (1824), 'A sketch of the geology, mineralogy and scenery of the region contiguous to the river Connecticut; with a geological map and drawings of organic remains; and occasional botanical notices', *American Journal of Science*, 7: 1–30.

—— (1833), *Report on the geology, mineralogy, botany, and zoology of Massachusetts*, Adams, Amherst, Mass.

HOLMES, A. (1965), *Principles of physical geology*, new and fully rev. edn., Nelson, London.

HOOKE, R. (1688), 'Lectures and discourses of earthquakes, and subterraneous eruptions: Explicating the causes of the rugged and uneven face of the Earth; and what reasons may be given for the frequent finding of shells and other sea and land petrified substances, scattered over the whole terrestrial superficies', published in 1705 by Richard Waller in his *The posthumous works of Robert Hooke: Containing his Cutlerian lectures, and other discourses, read at the meetings of the illustrious Royal Society* (see Waller 1705).

HOOYKAAS, R. (1970), 'Catastrophism in geology, its scientific character in relation to actualism and uniformitarianism', *Koninklijke Nederlandse Akademie van Wetenschappen, afd. Letterkunde, Med.*, NS, 33: 271–317.

HOPKINS, W. (1838), 'Researches in physical geology', *Transactions of the Cambridge Philosophical Society*, 6: 1–84.

—— (1842), 'On the elevation and denudation of the district of the lakes of Cumberland and Westmoreland', *Proceedings of the Geological Society of London*, 3: 757–66.

—— (1844), 'On the transport of erratic blocks', *Transactions of the Cambridge Philosophical Society*, 8: 220–40.

—— (1848), 'On the elevation and denudation of the district of the lakes of Cumberland and Westmoreland', *Quarterly Journal of the Geological Society of London*, 4: 70–98.

HOWORTH, H. H. (1887), *The mammoth and the Flood: An attempt to confront the theory of uniformity with the facts of recent geology*, Sampson Low, Marston, London.

—— (1893), *The glacial nightmare and the Flood: A second appeal to common sense from the extravagance of some recent geology*, 2 vols., Sampson Low, Marston, London.

—— (1905), *Ice or water: Another appeal to induction from the scholastic methods of modern geology*, 3 vols., Longmans, London.

HOYT, W. G. (1987), *Coon Mountain controversies: Meteor crater and the development of the impact theory*, University of Arizona Press, Tucson, Ariz.

HUGGETT, R. J. (1988), 'Terrestrial catastrophism: causes and effects', *Progress in Physical Geography*, 12: 509–32.

—— (1989), 'Superwaves and superfloods: The bombardment hypothesis and geomorphology', *Earth Surface Processes and Landforms*, 14: July issue.

—— (in preparation), *Climate and Earth history*, Springer Series in Physical Environment, Springer, Heidelberg.

HUTTON, J. (1785), 'Abstract of a dissertation read in the Royal Society of Edinburgh, upon the seventh of March, and fourth of April, M,DDC,LXXXC, concerning the system of the Earth, its duration, and stability', reprinted in abstract form in C. C. Albritton (ed.), *Philosophy of geohistory: 1785–1970*, Dowden, Hutchinson, and Ross, Stroudsberg, Pa., pp. 24–52.

—— (1788), 'Theory of the Earth', *Transactions of the Royal Society of Edinburgh*, 1: 209–304.

—— (1795), *Theory of the Earth*, 2 vols., William Creech, Edinburgh.

IMBRIE, J., and IMBRIE, K. P. (1986), *Ice ages: Solving the mystery*, Harvard University Press, Cambridge, Mass.

JAMESON, R. (1808), *Elements of Geognosy: Being Vol. III and Part II of the System of mineralogy*, William Blackwood, Edinburgh and Longman, Hurst, Rees, and Orme, London. [See Jameson (1976), *The Wernerian theory of the Neptunian origin of rocks: A facsimile reprint of Elements of geognosy, 1808 by Robert Jameson*, with an introduction by Jessie M. Sweet, Contributions to the History of Geology, vol. 9, Hafner Press, New York and Collier Macmillan, London.]

JAMIESON, T. F. (1865), 'On the history of the last geological changes in Scotland', *Quarterly Journal of the Geological Society of London*, 21: 161–203.

JANSA, L. F., and PE-PIPER, G. (1987), 'Identification of an underwater extraterrestrial impact crater', *Nature*, 327: 612–14.

JARRETT, R. D., and MALDE, H. E. (1987), 'Paleodischarge of the late Pleistocene Bonneville Flood, Snake River, Idaho, computed from new evidence', *Bulletin of the Geological Society of America*, 99: 127–34.

JOHNSON, J. G. (1971), 'Timing and coordination of orogenic, epeirogenic, and eustatic movements', *Bulletin of the Geological Society of America*, 82: 3263–98.

JONES, D. K. C. (1981), *Southeast and southern England* (The geomorphology of the British Isles series), Methuen, London.

JUKES, J. B. (1853), *Popular physical geology*, London.

—— (1862*a*), *The student's manual of geology*, new edn., Adam and Charles Black, Edinburgh.

—— (1862*b*), 'On the mode of formation of some river-valleys in the south of Ireland', *Quarterly Journal of the Geological Society of London*, 18: 378–403.

JURDY, D. M. (1981), 'True polar wander', *Tectonophysics*, 74: 1–16.

—— (1983), 'Early Tertiary subduction zones and hot spots', *Journal of Geophysical Research*, 86: 6395–402.

—— and VAN DER VOO, R. (1975), 'True polar wander since the Early Cretaceous', *Science*, 187: 1193–6.

KALM, P. (1937), *Peter Kalm's travels in North America*, the English version of 1770 by J. R. Forster. Rev. from the original Swedish and ed. Adolph B. Benson, with a trans. of new material from Kalm's diary notes, 2 vols., Wilson-Erickson, New York.

KEILL, J. (1698), *An examination of Dr. Burnet's theory of the Earth: Together with some remarks on Mr. Whiston's new theory of the Earth*, Theater, Oxford.

KELLY, S. (1969), 'Theories of the Earth in Renaissance cosmologies', in C. J. Schneer (ed.), *Toward a history of geology*, MIT Press, Cambridge, Mass., pp. 214–25.

KIRWAN, R. (1793), 'Examination of the supposed igneous origin of stony substances', *Transactions of the Royal Irish Academy*, 5: 51–81.

—— (1797), 'On the primitive state of the globe and its subsequent catastrophe', *Transactions of the Royal Irish Academy*, 6: 233–308.

—— (1799), *Geological essays*, D. Bremner, London.

KREICHGAUER, D. (1902), *Die Aquatorfrage in der Geologie*, 2nd edn., Missionsdruckerei, Steyl.

KURTÉN, B. (1986), *How to deep-freeze a mammoth*, based on a translation by E. J. Friis of the original work, *Hur man fryser in ein mammut*, Columbia Press, New York.

KYTE, F. T., LEI ZHOU, and WASSON, J. T. (1988), 'New evidence on the size and possible effects of a late Pliocene oceanic asteroid impact', *Science*, 241: 63–5.

LAMARCK, J. B. P. A. DE MONET, CHEVALIER DE (1802), *Hydrogéologie, ou recherches sur l'influence qu'ont les eaux sur la surface du globe terrestre; sur les causes de l'existence du bassin des mers, de son déplacement et de son transport successif sur les différentes points de la surface de ce globe; enfin sur les changemens que les corps vivans exercent sur la nature et l'état de cette surface*, Paris. [See also Lamarck (1964), *Hydro-geology*, trans. and ed. Albert Carozzi, University of Illinois Press, Urbana, Ill.]

LA PEYRÈRE, I. DE (1655), *Praeadamitae, sive exercitatio super versibus duodecimo, decimotertio, et decimoquarto, capitis quinti epistolae D.*

Pauli ad Romanos: Quibus inducuntur primi homines ante Adamum conditi. (Systema theologicum ex Praeadamitarum hypothesi. Pars prima.), Amsterdam. [Trans. in 1656 as *Men before Adam, or a discourse upon the twelfth, thirteenth, and fourteenth verses of the fifth chapter of the epistle of the Apostle Paul to the Romans: By which are prov'd that the first men were created before Adam. (A theological systeme upon that presupposition that men were before Adam. The first part.)*, London.]

LAPLACE, P. S. DE (1796), *Exposition du système du Monde*, 2 vols., Paris. [See also Laplace (1809), *The system of the World*, trans. J. Pond, London.]

LARSON, R. L., and PITMAN III, W. C. (1972), 'World-wide correlation of Mesozoic magnetic anomalies, and its implications', *Bulletin of the Geological Society of America*, 83: 3645–62.

LEHMANN, J. G. (1756), *Versuch einer Geschichte von Flötz-Gebürgen, betrefend deren Entstehung, Lage, darinne befindliche, Metallen, Mineralien und Fossilien*, Berlin.

LEIBNITZ, G. W. VON (1749), *Protogaea, sive de prima facie telluris et antiquissimae historiae vestigiis in ipsis naturae monumentis dissertatio, ex schedis manuscriptis viri illustris in lucem edita a Christiano Ludovico Scheidio*, Göttingen.

LE ROY, LOYS (1594), *Of the interchangeable course or variety of things in the whole world*, trans. R. A., Charles Yetsweirt, London.

LIPSCHUTZ, M. E., and ANDERS, E. (1961), 'The record in the meteorites—IV: Origin of diamonds in iron meteorites', *Geochimica et Cosmochimica Acta*, 24: 83–105.

LÖFFELHOLZ VON COLBERG, C. F. (1886), *Die Drehung der Erdkruste in geologischen Zeiträumen: Ein neuer geologisch-astronomische Hypothese*, J. A. Finsterlin, Munich.

LUBBOCK, J. (1848), 'On change of climate resulting from a change in the Earth's axis of rotation', *The Quarterly Journal of the Geological Society of London*, 4: 4–7.

LUTHER, M. (1958), *Luther's commentary on Genesis*, vol. i, trans. J. Theodore Muller, Eerdmans, Grand Rapids, Mich.

LYELL, C. (1830–3), *Principles of geology, being an attempt to explain the former changes of the Earth's surface, by reference to causes now in operation*, 3 vols., John Murray, London.

—— (1834), *Principles of geology, being an inquiry how far the former changes of the Earth's surface are referable to causes now in operation*, 3rd edn., 4 vols., John Murray, London. (This is the edition from which the quotations in the text are taken.)

—— (1874), *The student's elements of geology*, 2nd edn., John Murray, London.

LYTTLETON, R. A. (1982), *The Earth and its mountains*, Wiley, Chichester.

McCALL, G. J. H . (ed.) (1979), *Astroblemes—cryptoexplosion structures*, Benchmark Papers in Geology, vol. 50, Dowden, Hutchinson, and Ross, Stroudsberg, Pa.

MACCULLOCH, J. (1817), 'On the parallel roads of Glen Roy', *Transactions of the Geological Society of London*, 4: 314–92.

MACKINTOSH, D. (1869), *The scenery of England and Wales, its character and origin: being an attempt to trace the nature of the geological causes, especially denudation, by which the physical features of the country have been produced: Founded on the results of many years' personal observations, and illustrated by eighty-six woodcuts, including sections, and views of scenery from original sketches or from photographs*, Longmans, Green, London.

MAILLETT, B. DE (TELLIAMED) (1748) *Telliamed ou entretiens, sur la diminution de la mer, d'un philosophe Indien avec un missionaire Français*, Amsterdam. [See De Maillet (1968), *Telliamed or conversations between an Indian philosopher and a French missionary on the diminution of the sea*, trans. and ed. Albert V. Carozzi, University of Illinois Press, Urbana, Ill.]

MAJOLI, S. (1597), *Dies caniculares seu colloquia tria et viginti; Quibis pleraque naturae admiranda quae aut in aethere fiunt, aut in Europa, Asia atque Africa, quin etiam in ipso orbe nouo, et apud omnes antipodes sunt recensentur ordine quem sequens pagina indicabit*, A. Zanetti, Rome.

MALDE, H. E. (1968), 'The catastrophic late Pleistocene Bonneville Flood in the Snake River Plain, Idaho', *US Geological Survey Professional Paper* 596.

MANDELKEHR, M. M. (1983), 'An integrated model for an Earthwide event at 2300 BC. Part I: The archaeological evidence', *Chronology and Catastrophism Review*, 3: 77–95.

—— (1987), 'An integrated model for an Earthwide event at 2300 BC. Part II: The climatological evidence', *Chronology and Catastrophism Review*, 9: 34–44.

MARK, K. (1987), *Meteorite craters*, University of Arizona Press, Tucson, Ariz.

MARRIOTT, R. A. (1914), *The change in the climate and its cause: Giving the date of the last ice age based on a recent astronomical discovery and geological research*, E. Marlborough, London.

MATHER, K. F., and MASON, S. L. (1939), *A source book in geology*, McGraw-Hill, New York.

MAXWELL, J. C. (1890), 'On a dynamical top', in W. D. Niven (ed.), *The scientific papers of James Clerk Maxwell*, vol. i, Cambridge University Press, Cambridge.

MENARD, H. W. (1964), *Marine geology of the Pacific*, McGraw-Hill, New York.

—— (1969), 'Elevation and subsidence of oceanic crust', *Earth and Planetary Science Letters*, 6: 275–84.

MERRILL, G. P. (1924), *The first one hundred years of American geology*, Yale University Press, New Haven, Conn., and Oxford University Press, London.

MITCHELL, ELISHA (1827), *On the character and origin of the low country of North Carolina*, University of North Carolina, Chapel Hill.

MITCHELL, S. L. (1818), 'Observations on the geology of North America; illustrated by the description of various organic remains, found in that part of the world', in Cuvier's *Essay on the theory of the Earth*, Kirk and Mercein, New York, pp. 321–428.

MOFFATT, A. J., and CATT, J. A. (1986), 'A re-examination of the evidence for a Plio-Pleistocene marine transgression on the Chiltern Hills. III. Deposits', *Earth Surface Processes and Landforms*, 11: 233–47.

MORGAN, W. J. (1983), 'Hotspot tracks and the early rifting of the Atlantic', *Tectonophysics*, 94: 123–39.

MÖRNER, N.-A. (1987), 'Models of global sea-level changes', in M. J. Tooley and I. Shennan (eds.), *Sea-level changes*, Basil Blackwell, Oxford, pp. 332–55.

MORO, A. L. (1740), *Dei crostacei e degli altri marini corpi che si trovano sui monti libre due*, Ceremia, Venice.

MURCHISON, R. I. (1867), *Siluria*, 4th edn., John Murray, London.

—— (1869), 'Address of the President', *Journal of the Royal Geographical Society*, 39: 135–94.

NANCE, R. D., WORSLEY, T. R., and MOODY, J. B. (1988), 'The supercontinent cycle', *Scientific American*, 259: 44–51.

NEWBERRY, J. S. (1861), 'Geological report of the Colorado exploring expedition, 1857–58', in J. C. Ives, *Report upon the Colorado River of the West,* US 36th Congress, 1st Session, Senate and House Executive Document 90, Part III.

NEWTON, I. (1729), *The mathematical principles of natural philosophy*, trans. Andrew Motte, 2 vols., Benjamin Motte, London.

NININGER, H. H. (1942), 'Cataclysm and evolution', *Popular Astronomy*, 50: 270–2.

—— (1956), *Arizona's meteorite crater*, American Meteorite Laboratory, Denver, Colo.

ORBIGNY, A. D. D' (1840–89), *Paléontologie française: Description zoologique et géologique de tous les animaux mollusques et rayonnés fossiles de la France . . . par Alcide d'Orbigny*, A. Bertrand, V. Masson, Paris.

OVID, (edn. 1955), *Metamorphoses*, trans. Mary M. Innes, Penguin Books, Harmondsworth.

PAGE, D. (1865), *Handbook of geological terms, geology and physical geography*, William Blackwood, Edinburgh and London.

PALISSY, B. (1580), *Discours admirables de la nature des eau et founteines tant naturelles qu'artificelles, des metaus, des sels et salines, des pierres, des terrers*, etc., Martin le Jeune, Paris. [See also Palissy (1957), *The admirable discourses of Bernard Palissy*, trans. and ed. A. La Roque, University of Illinois Press, Urbana, Ill.]

PALLAS, P. S. (1771), *Observations sur la formation des montagnes et les changements arrivés au globe, particulièrement de l'Empire Russe*, St Petersburg.

PARACELSUS, P. A. (1951), *Selected writings*, trans. Norbert Guterman (Bollinger Series 28), Pantheon Books, New York.

PARTRIDGE, J., and BAKER, V. R. (1987), 'Palaeoflood hydrology of the Salt River, Arizona', *Earth Surface Processes and Landforms*, 12: 109–25.

PENN, G. (1828), *Conversations on geology: Comprising a familiar explanation of the Huttonian and Wernerian systems; the Mosaic geology, as explained by Mr. Granville Penn; and the late discoveries of Professor Buckland, Humboldt, Dr. Macculloch, and others*, Samuel Maunder, London.

PHILLIPS, J. (1853), *The rivers, mountains, and sea-coast of Yorkshire: With essays on the climate, scenery, and ancient inhabitants in the county*, John Murray, London.

PITMAN III, W. C. (1978), 'Relationship between eustacy and stratigraphic sequences of passive margins', *Bulletin of the Geological Society of America*, 89: 1389–403.

PITTY, A. F. (1983), *The nature of geomorphology*, Methuen, London.

PLATO, (edn. 1971), *Timaeus* and *Critias*, trans. with an introduction and an appendix on *Atlantis* by Desmond Lee, Penguin Books, Harmondsworth.

PLATTES, G. (1639a), *A discovery of infinite treasure, hidden since the world's beginning: Whereunto all men, of what degree soever, are friendly invited to be sharers with the discoverer, G. P.*, George Hutton for I. Leggatt, London.

—— (1639b), *A discovery of subterraneall treasure: Viz. of all manner of mines and minerals from the gold to the coale, with plain directions and rules for finding them in all kingdoms and countries*, I. Okes for J. Emery, London.

PLAYFAIR, J. (1802), *Illustrations of the Huttonian theory of the Earth*, Cadell and Davies, London, and Creech, Edinburgh.

—— (1811), 'Review of Transactions of the Geological Society of London', *Edinburgh Review*, 19: 207–29.

PORTER, R. (1977), *The making of geology: Earth science in Britain 1660–1815*, Cambridge University Press, Cambridge.

POWNALL, T. (1776), *A topographical description of such parts of North America as are contained in the (annexed) map of the middle British*

Colonies in North America, J. Almon, London. [See also Pownall (1949), *A topographical description of the dominions of the United States of America: Being a revised and enlarged edition of A topographical description of . . . North America*, ed. Lois Mulkearn, University of Pittsburgh Press, Pittsburgh, Pa.]

PRESTWICH, J. (1862–3*a*), 'Theoretical considerations on the conditions under which the drift deposits containing the remains of extinct mammalia and flint implements were accumulated; and on their geological age', *Proceedings of the Royal Society*, 12: 38–52.

—— (1862–3*b*), 'On the loess of the valleys of the south of England and of the Somme and the Seine', *Proceedings of the Royal Society*, 12: 170–3.

—— (1863–4), 'On some further evidence bearing on the excavation of the valley of the Somme by river action', *Proceedings of the Royal Society*, 13: 135–7.

—— (1864), 'Theoretical considerations on the conditions under which the (drift) deposits containing the remains of extinct mammalia and flint implements were accumulated, and on their geological age', *Philosophical Transactions of the Royal Society*, 154: 247–309.

QUIRINI, J. (1676), *De testaceis fossilibus musaei Septalliani et Jacobi Grandii de veritate diluvii universalis, et testaceorum, quae procul a mari reperiuntur generatione*, Valuasensis, Venice.

RAMSAY, A. C. (1846), 'On the denudation of South Wales and the adjacent counties of England', *Memoirs of the Geological Survey of Great Britain*, HMSO London, 1: 297–335.

RAUP, D. M. (1982), 'Biogeographic extinction: A feasibility test', in L. T. Silver and P. H. Schultz (eds.), *Geological implications of impacts of large asteroids and comets on the Earth*, Geological Society of America Special Paper 190, pp. 277–81.

RAY, J. (1691), *The wisdom of God manifested in the works of the Creation*, Samuel Smith, London.

—— (1692), *Miscellaneous discourses concerning the dissolution and changes of the world*, Samuel Smith, London.

—— (1693), *Three physico-theological discourses, containing I. The primitive chaos, and creation of the world. II. The general Deluge, its causes and effects. III. The dissolution of the world, and future conflagration. Wherein are largely discussed the production and use of mountains; the original of fountains, of formed stones, and sea-fishes bones and shells found in the Earth; the effects of particular floods and inundations of the sea; the eruptions of volcano's; the nature and causes of earthquakes: with an historical account of those two remarkable ones in Jamaica and England. With practical inferences*, Samuel Smith, London.

RAZUMOVSKY, G. DE (1789), *Histoire naturelle du Jorat et de ses environs;*

et celle trois lacs Neufchatel, Morat et Bienne; précédées d'un essai sur le climat, les productions . . . de la partie du Pays de Vand . . . qui entre dans le plan de cette ouvrage, 2 vols., J. Mourer, Lausanne.

REIBISCH, P. (1901), 'Ein Gestaltungsprinzip der Erde', *Jahresbericht des Vereins für Erdkunde zu Dresden*, 27: 105–24.

REID, C. (1915), 'Ancient rivers of Bournemouth', *Proceedings of the Bournemouth Natural Science Society*, 7: 73–82.

ROGERS, H. D. (1835), 'On the falls of the Niagara and the reasonings of some authors respecting them', *American Journal of Science*, 27: 326–35.

ROSENBERG, D. (1986), *World mythology: An anthology of the great myths and epics*, Harrap, London.

RUDWICK, M. J. S. (1972), *The meaning of fossils*, MacDonald, London.

RUPKE, N. A. (1983), *The great chain of history: William Buckland and the English School of Geology (1814–1849)*, Clarendon Press, Oxford.

RUSSEL, K. L. (1968), 'Ocean ridges and eustatic changes in sea level', *Nature*, 261: 680–2.

SAINTE-CLAIRE DEVILLE, C. (1978), *Coup d'œil historique sur la géologie et sur les travaux d'Élie de Beaumont: Leçons professées au Collège de France (mai–juillet 1875), par Charles Sainte-Claire Deville*, G. Masson, Paris.

SANDARS, N. K. (1960), 'Introduction' to an English version of *The epic of Gilgamesh*, Penguin, Harmondsworth.

SAULL, W. D. (1848), 'An elucidation of the successive changes of temperature and the levels of oceanic waters upon the Earth's surface, in harmony with geological evidences', *The Quarterly Journal of the Geological Society of London*, 4: 7.

SAUSSURE, H.-B. DE (1779–96), *Voyages dans les Alpes, précédés d'un essai sur l'histoire naturelle dans les environs Genève*, 4 vols., S. Fauché, Neuchâtel.

SCHEUCHZER, J. J. (1708), *Piscium querelae et vindicae*, Gesner, Zurich.

SCILLA, A. (1670), *La vana speculazione disingannata dal senso, lettera risponsiva circa i corpi marini che petrificati si trovano in varii luoghi terrestri*, A. Colicchia, Naples.

SCOULER, J. (1836–7), 'On certain elevated hills of gravel containing marine shells in the vicinity of Dublin', *Proceedings of the Geological Society of London*, 2: 435–7.

SCROPE, G. J. P. (1825), *Considerations on volcanos*, Phillips and Yard, London.

SEDGWICK, A. (1825), 'On the origin of alluvial and diluvial formations', *Annals of Philosophy*, 9: 241–57.

—— (1834), 'Address to the Geological Society, delivered on the evening of the 18th of February 1831, by the Rev. Professor Sedgwick, M.A. F.R.S. &c. on retiring from the President's chair', *Proceedings of the Geological Society of London*, 1: 281–316.

SHARPE, D. (1856), 'On the elevation of the Alps, with notices of the heights at which the sea has left traces of its action on their sides', *Quarterly Journal of the Geological Society of London*, 12: 102–23.

SHOEMAKER, E. M. (1963), 'Impact mechanisms at Meteor Crater, Arizona', in B. M. Middlehurst and G. P. Kuiper (eds.), *The Moon, meteorites and comets*, University of Chicago Press, Chicago, pp. 301–6.

—— (1983), 'Asteroid and comet bombardment of the Earth', *Annual Reviews of Earth and Planetary Sciences*, 11: 461–94.

—— (1984), 'Large body impacts through geologic time', in H. D. Holland and A. F. Trendall (eds.), *Patterns of change in Earth evolution*, Springer-Verlag, Berlin, pp. 15–40.

SILLIMAN, B. (1821), 'Notice of the "Geological essays, or an inquiry into some of the geological phenomena, to be found in various parts of America and elsewhere—by Horace H. Hayden, Esq. member of the American Geological Society," &c, &c.', *American Journal of Science*, 3: 47–57.

SIMOENS, G. (1937), *La théorie de l'évolution cataclysmique et de l'évolution alternante*, Dunod, Paris.

SIMPSON, G. G. (1970), 'Uniformitarianism: An inquiry into principle, theory, and method in geohistory and biohistory', in M. K. Hecht and W. C. Steere (eds.), *Essays in evolution and genetics in honor of Theodosius Dobzhansky*, Appleton-Century-Crofts, New York, pp. 43–96.

SINGER, D. W. (1950), *Giordano Bruno, his life and thought: With annotated translation of his work on the infinite universe and worlds*, Henry Schuman, New York.

SLABINSKI, V. J. (1981), 'A dynamical objection to the inversion of the Earth on its spin axis', *Journal of Physics A: Mathematical and General*, 14: 2503–7.

—— (1982), 'Dr. Victor J. Slabinski comments', *Kronos*, 7: 94–6.

SOULAVIE, J. L. G. (1780–4), *Histoire naturelle de la France Méridionale*, 8 vols., Paris.

STEDINGER, J. R., and BAKER, V. R. (1987), 'Surface water hydrology: Historical and paleoflood information', *Reviews of Geophysics*, 25: 119–24.

STENO, N. (1669), *Nicolai Stenonis de solido intra solidum naturaliter contento dissertationis produmus: Ad serenissimum Ferdinandum II. Magnum Etruriae Ducem*, Ex typographia sub signo stellae, Florence. [See Steno (1916), *The prodomus of Nicolaus Steno's dissertation concerning a solid body enclosed by process of nature within a solid*, an English version with an introduction and explanatory notes by John Garrett Winter, Macmillan, New York and London.]

STEVENSON, T. (1849) 'Account of experiments upon the force of the waves of the Atlantic and German Oceans', *Transactions of the Royal Society of Edinburgh*, 16: 23–32.

STEVIN, SIMON (1634), *Les Œuvres mathematiques de Simon Stevin*, revised and corrected by Albert Girard, B. & A. Elsevier, Leiden.

STILLINGFLEET, E. (1662), *Origines sacrae, or a rational account on the Christian faith, as to the truth and divine authority of the scriptures, and the matters therein contained*, Henry Mortlock, London.

—— (1836), *Origines sacrae: Or, a rational account of the grounds of natural and revealed religion: Together with a letter to a deist*, 2 vols., Oxford University Press, Oxford.

STRABO (1966–70), *The geography of Strabo*, trans. Horace Leonard Jones, 8 vols. (The Loeb Classical Library vols. 49–50, 182, 196, 211, 223, 241, 267), Heinemann, London.

STRAKHOV, N. M. (1949), 'Periodicity and irreversibility of evolution of sedimentation in the Earth's history', *Izvestiya Akademii Nauk SSSR, Seriya Geologiya*, 6: 70–111 [in Russian].

STRELITZ, R. A. (1979), Meteorite impact in the ocean', *Proceedings of the Tenth Lunar and Planetary Science Conference*, 3: 2799–813.

SUESS, E. (1885–1909), *Das Anlitz der Erde*, 5 vols., Freytag, Vienna.

—— (1904–24), *The face of the Earth*, 5 vols., trans. H. B. C. Sollas, Clarendon Press, Oxford.

SULIVAN, R. J. (1794), *A view of nature, in letters to a traveller among the Alps with reflections on atheistical philosophy, now exemplified in France*, 6 vols., T. Becket, London.

SULLIVAN, W. (1974), *Continents in motion*, McGraw-Hill, New York.

TARGIONI-TOZZETTI, G. (1751–4), *Relazioni d'alcuni viaggi fatti in diverse parti della Toscana, per osservare le produzioni naturali e gli antichi monumenti di essa*, Stamp. Imperiale, Florence.

THOMSON, W. (1876), 'Sectional address', *Nature*, 14: 429.

TOWNSON, R. (1794), *Philosophy of mineralogy*, printed by the author, sold by John White, London.

TRIMMER, J. (1834), 'On the diluvial deposits of Caernarvonshire between the Snowdon chain of hills and the Menai strait, and on the discovery of marine shells in diluvial sand and gravel on the summit of Moel Tryfane, near Caernarvon, 1000 ft. above the level of the sea', *Proceedings of the Geological Society of London*, 1: 331–2.

—— (1854), 'On pipes and furrows in calcareous and non-calcareous strata', *Quarterly Journal of the Geological Society of London*, 10: 231–40.

TWISDEN, J. F. (1878), 'On possible displacements of the Earth's axis of figure produced by elevations and depressions of her surface', *Quarterly Journal of the Geological Society of London*, 34: 35–48.

TYLOR, A. (1866), 'Remarks on the interval of time which has passed between the formation of the upper and lower valley-gravels of part of

England and France; with notes on the character of holes bored in rocks by mollusca', *Quarterly Journal of the Geological Society of London*, 22: 463–8.

—— (1868), 'On the Amiens gravel', *Quarterly Journal of the Geological Society of London*, 24: 103–25.

—— (1869), 'On the formation of deltas: And on the evidence and cause of great changes in the sea-level during the glacial period', *Quarterly Journal of the Geological Society of London*, 25: 7–11.

—— (1872), 'On the formation of deltas: And on the evidence and cause of great changes of sea-level during the glacial period', *Geological Magazine*, 9: 392–9, 485–501.

UMBGROVE, J. H. F. (1947), *The pulse of the Earth*, 2nd edn., Nijhoff, The Hague.

VAIL, P. R., MITCHUM, R. M. JUN., and THOMPSON III, S. (1977), 'Seismic stratigraphy and global changes of sea level, Part 4: Global cycles of relative changes of sea level', *Memoirs of the American Association of Petroleum Geologists*, 26: 83–97.

VALENTINE, J. W., and MOORES, E. M. (1972), 'Global tectonics and the fossil record', *Journal of Geology*, 80: 167–84.

VALLISNIERI, A. (1721), *Dei corpi marini, che sui monti si trovano; della loro origine; et dello stato del mondo avanti il diluvio, nel diluvio, e dopo il diluvio* etc., Venice.

VELCURIO, J. (1588), *Ioannis Velcurionis commentariorum libri iiii, in universam Aristotelis physicem nunc recens summa fide exactaque diligentia castigati et excusi*, George Bishop, London.

VELIKOVSKY, I. (1950), *Worlds in collision*, Doubleday, Garden City, New York.

—— (1952), *Ages in chaos*, Doubleday, Garden City, New York.

—— (1955), *Earth in upheaval*, Doubleday, Garden City, New York.

VOSSIUS, I. (1659), *Dissertatio de vera aetate mundi, qua ostenditur mundi tempus annis minimum 1440 vulgarem aeram anticipare*, Adriani Vlacq, The Hague.

WALLER, R. (1705), *The posthumous works of Robert Hooke: Containing his Cutlerian lectures, and other discourses, read at the meetings of the illustrious Royal Society*, Richard Waller for the Royal Society, London. [A 2nd facsimile edn. of this work, with a new introduction by T. M. Brown of Princeton University, was published by Frank Cass, London, 1978.]

WARLOW, P. (1978), 'Geomagnetic reversals?', *Journal of Physics A: Mathematical and General*, 11: 2107–30.

—— (1982), *The reversing Earth*, Dent, London.

—— (1987), 'Return to the tippe top', *Chronology and Catastrophism Review*, 9: 2–13.

WEGENER, A. L. (1915), *Die Entstehung der Kontinente und Ozeane*, Friedrich Vieweg und Sohn, Brunswick.

—— (1929), *Die Entstehung der Kontinente und Ozeane*, 4th edn., Friedrich Vieweg und Sohn, Brunswick. [A translation by J. Biram of this, the 4th, rev. edn. of the book, was published by Methuen, London, 1966.]

WERNER, A. G. (1787), *Kurze Klassifikation und Beschreibung der verschiedener Gebirgsarten*, published by a friend, Dresden. [See also Werner, (1971), *Short classification and description of the various rocks*, trans. A. Ospovat, Hafner, New York.]

WEYER, E. M. (1978), 'Pole movement and sea levels', *Nature*, 273: 18–21.

WHEWELL, W. (1847), 'On the wave of translation in connexion with the Northern Drift', *Quarterly Journal of the Geological Society of London*, 3: 227–32.

WHISTON, W. (1696), *A new theory of the Earth, from its original, to the consummation of all things: Wherein the Creation of the world in six days, the universal deluge, and the general conflagration, as laid down in the Holy Scriptures, are shewn to be perfectly agreeable to reason and philosophy. With a large introductory discourse concerning the genuine nature, stile, and extent of the* Mosaick *History of the Creation*, Benjamin Tooke, London.

—— (1717), *Astronomical principles of religion, natural and revealed*. J. Senex and W. Taylor, London. [See also Whiston [1983] *Astronomical principles of religion, natural and reveal'd*, with an introduction by James E. Force, Georg Orms Verlag, Hildesheim, Zurich, New York.]

WHITE, G. W. (1956), 'Lewis Evans (1700–56): a scientist in Colonial America', *Nature*, 177: 1055–6.

WHITE, J. (1986), *Pole shift: Predictions and prophecies of the ultimate disaster*, A.R.E. Press, Virginia Beach, Va.

WHITEHURST, J. (1778), *An inquiry into the original state and formation of the Earth; deduced from facts and the laws of nature: To which is added an appendix, containing some general observations on the strata in Derbyshire: With sections of them, representing their arrangement, affinities, and the mutations they have suffered at different periods of time, intended to illustrate the preceding enquiries, and as a specimen of subterraneous geography*, printed for the author by J. Cooper, London.

WILLIAMS, G. E. (1972), 'Geological evidence relating to the origin and secular rotation of the Solar System', *Modern Geology*, 3: 165–81.

—— (ed.) (1981), *Megacycles: Long-term episodicity in Earth and planetary history*, Benchmark Papers in Geology Vol. 57, Dowden, Hutchinson, and Ross, Stroudsberg, Pa.

Williams, J. (1789), *The natural history of the mineral kingdom, relative to the strata of coal, mineral veins, and the prevailing strata of the globe*, 2 vols., Edinburgh.

—— (1810), *The natural history of the mineral kingdom, relative to the strata of coal, mineral veins, and the prevailing strata of the globe. With an appendix, containing a more extended view of mineralogy and geology*, 2 vols., Bell & Bradfute, and W. Laing, Edinburgh; Longman, Hurst, Rees, and Orme; and J. White, London.

Winter, J. G. (1916), *The prodomus of Nicolaus Steno's dissertation concerning a solid body enclosed by process of nature within a solid*, an English version with an introduction and explanatory notes, University of Michigan Humanistic Series, Vol. xi, Contributions to the History of Science, Part II, Macmillan, New York and London.

Wise, D. U. (1974), 'Continental margins, freeboard and the volumes of continents and oceans through time', in C. A. Burk and C. L. Brake (eds.), *The geology of continental margins*, Springer-Verlag, Berlin, pp. 45–58.

Wood, S. V. Jun. (1864), 'On the formation of river- and other valleys of the east of England', *Philosophical Magazine*, 4th series, 27: 180–90.

—— (1865), 'Letter under Scientific Correspondence', *The Reader: A Review of Literature, Science, and Art*, 6: 297.

Woodward, J. (1695), *An essay towards a natural history of the Earth: and terrestrial bodies, especially minerals: As also of the sea, rivers, and springs: With an account of the universal deluge: And of the effects that it had upon the Earth*, Richard Wilkin, London.

Worsley, T. R., Nance, R. D., and Moody, J. B. (1984), 'Global tectonics and eustacy for the past 2 billion years', *Marine Geology*, 58: 373–400.

Zittel, K. A. von (1901), *History of geology and palaeontology to the end of the nineteenth century*, trans. Maria M. Olgivie-Gordon, Walter Scott, London.

Index